Security Orchestration, Automation, and Response

for Security Analysts

Learn the secrets of SOAR to improve MTTA and MTTR and strengthen your organization's security posture

Benjamin Kovacevic

BIRMINGHAM—MUMBAI

Security Orchestration, Automation, and Response for Security Analysts

Group Product Manager: Pavan Ramchandani
Publishing Product Manager: Prachi Sawant
Senior Editor: Athikho Sapuni Rishana
Technical Editor: Arjun Varma
Copy Editor: Safis Editing
Project Coordinator: Ashwin Kharwa
Proofreader: Safis Editing
Indexer: Hemangini Bari
Production Designer: Nilesh Mohite
Marketing Coordinator: Shruthi Shetty, Marylou De Mello
Business Development Executive: Prathamesh Walse

First published: July 2023
Production reference: 1230623

Published by Packt Publishing Ltd.
Livery Place
35 Livery Street
Birmingham
B3 2PB, UK.

978-1-80324-291-0

www.packtpub.com

Foreword

In today's threat landscape, it's very important to respond to incidents and alerts in a timely manner. An organization's **Security Operations Center (SOC)** can become overwhelmed if too many alerts are generated and there are not enough SOC analysts to triage them, or skilled cybersecurity workers to fill the positions needed to respond. This is why an automated response to security incidents is a must. **Security Orchestration, Automation, and Response (SOAR)** is the answer to an organization's SOC overcoming these challenges. SOAR can be used to reduce the amount of alerts that need investigation and triage, to automate parts of normal investigations and save SOC analysts' time, and more importantly, to automate remediation, leading to quick actions to resolve security incidents.

In this book, you will learn about security orchestration, automation, and response in depth, both in theory and principle, using real playbook examples to automate a response. You will learn about the tools available and the various methods to implement them as a partial or complete security incident response. Benjamin is a skilled professional and expert in SOAR, helping customers implement it and creating samples, shared through open source to help organizations enable their security automation.

Nicholas DiCola,

Vice President of Customers

Zero Networks

Contributors

About the author

Benjamin Kovacevic is a cybersecurity enthusiast with hands-on experience with Microsoft XDR and SIEM platforms. Currently working with Microsoft Sentinel as a product manager, he focuses on the SOAR component of Microsoft Sentinel and works on new capabilities that help SOCs improve their investigations and responses. Benjamin constantly works to improve his knowledge about cybersecurity and also shares his knowledge about Microsoft SOAR. He is the author of Microsoft Sentinel Automation training blog, as well as many other blog posts containing tips and tricks to get started quickly with Microsoft Sentinel Automation.

Benjamin is originally from Bosnia and Herzegovina, but he currently resides in Ireland with his wife and two sons.

I want to thank my wife, Dzenana, and my sons, Adi and Mak. Thank you for all the sacrifices you have made and for supporting me through this journey. Also, a big thanks to all the people who have made a big impact on my security journey!

About the reviewers

Guven Boyraz boasts over a decade of experience in the computer science and IT industry, specialising in cybersecurity and software product development. Throughout his career, he has provided cybersecurity consultancy services to a wide range of clients, including both enterprise-level customers and startups, primarily in London, UK. With a BSc in electrical and electronics engineering and several certifications in computer science, Boyraz has acquired a strong educational foundation. In addition to his consulting work, he has also made significant contributions as a trainer and speaker at numerous international conferences.

I truly believe all of us in the technical world and science are standing on the shoulders of giants. The giants for me are the open communities, such as OWASP, Linux Commuties, and GitHub, where access to information is unrestricted and people are interested in helping one another. I am deeply indebted to all the communities and the people running them. I am also thankful to my family for providing all the lifelong support and love.

Javier Soriano has more than 15 years of experience as an IT solutions architect. He has worked in multiple areas within the IT field, such as storage, virtualization, automation, and security. His current role is Senior Program Manager in the Security Engineering division at Microsoft, where he helps customers and partners implement and operate their security operations with Microsoft Sentinel.

Table of Contents

Part 2: SOAR Tools and Automation Hands-On Examples

6

Enriching Incidents Using Automation   131

7

Managing Incidents with Automation   211

8

Responding to Incidents Using Automation   255

9

Mastering Microsoft Sentinel Automation: Tips and Tricks   287

Preface

Hey everyone! In this book, we will cover the topic of **Security Orchestration, Automation, and Response (SOAR)**. SOAR is one of the main tools in **Security Operation Centers (SOCs)** as it provides a unique set of features that allows you to perform needed steps from incident creation to incident resolution.

There are four main elements in SOAR that we will focus on:

- Incident management

- Incident investigation

- Automation

- Reporting

We will cover all four elements in the book, starting with an overview of each element, and then showcase what each element looks like in tools such as Microsoft Sentinel, Splunk SOAR, and Google Chronicle SOAR.

As there is a lot of documentation on incident management and investigation from the perspective of incident management frameworks, the second part of the book will focus on automation with hands-on examples using Microsoft Sentinel.

I will provide step-by-step instructions on how to create, test, and utilize automation using Microsoft Sentinel based on personal experience working with Microsoft Sentinel automation, the documentation available, and experience gathered working with many different organizations adapting Microsoft Sentinel, and especially Microsoft Sentinel SOAR.

The reason why we are focusing on SOAR is that it is a growing market, and many organizations are starting to adapt SOAR tools for their day-to-day operations. This can be seen best from the perspective that each **Security Information and Event Management (SIEM)** vendor has either created its own SOAR solution, bought a SOAR solution and integrated it with its SIEM, or has close cooperation with an independent SOAR vendor.

Who this book is for

This book is mainly written for security professionals that don't have experience working with SOAR, but also for any IT professional that wants to learn more about incident management, investigation, and automation using SOAR.

This book can also be used for those who want to learn how to create, test, and utilize Microsoft Sentinel automation for day-to-day activities.

What this book covers

Chapter 1, *The Current State of Cybersecurity and the Role of SOAR*, is a general overview of cybersecurity, why traditional tools aren't enough in the fight, and how modern tools add value to a SOC. We will continue with the topic of SOAR, what it is, why it's one of the SOC analysts' "best friends," and how it can reduce the amount of time required to respond to incidents.

Chapter 2, *A Deep Dive into Incident Management and Investigation*, will focus on incident management and investigation, its importance, and some of the best approaches to incident management and investigation. This will include owner assignment, collaboration, modern tools, and lessons learned as one of the most critical aspects of incident investigation.

Chapter 3, *A Deep Dive into Automation and Reporting*, provides an overview of automation as one of the most significant elements of SOAR. We will cover automation as a SOC's best friend, why you should be using it, and what we can automate. In this chapter, we will go through reporting, as well as how it can help SOCs be more efficient.

Chapter 4, *Qucik Dig into SOAR Tools*, will go over the most known SOAR tools, how they look, and what options they have. In it, we will go through the importance of SOAR and how it changed the traditional SIEM space.

Chapter 5, *Introducing Microsoft Sentinel Automation*, will introduce all aspects of Microsoft Sentinel automation on a more profound level, as a continuation of the Microsoft Sentinel SOAR intro in the previous chapter. We will be explaining topics such as automation rules and playbooks and how to utilize them to fight the dark side.

Chapter 6, *Enriching Incidents Using Automation*, focuses on the first hands-on example, where we will show you how to utilize solutions such as VirusTotal to enrich incidents on creation/update. We will go over enrichment and how we can use it to improve the amount of time taken for initial triage from hours to minutes!

Chapter 7, *Managing Incidents with Automation*, will focus on incident management with automation, how to control false-positive/low-severity incidents, and user/SOC analyst inputs for faster incident resolution. MTTA and MTTR are the main SOC measurements, and proper automation will lower both of them.

Chapter 8, *Responding to Incidents Using Automation*, will focus on responding to the incident as one of the most critical automation scenarios. Examples include blocking the user, isolating the host, blocking the IP, resetting users' passwords, and so on. A fast response can isolate a bad actor in its initial stage, and with automation, this can be done as soon as the incident is created – with or without SOC analyst interaction.

Chapter 9, Mastering Microsoft Sentinel Automation: Tips and Tricks, will go over tips and tricks for using Microsoft Sentinel as an automation tool. We will demonstrate its power under the hood and how to utilize automation below the GUI. This will include the options for automatically adding "hidden" elements, functions for better content control, and everything about HTTP action.

To get the most out of this book

A basic understanding of Microsoft Azure, SIEM, and JSON would be beneficial, but we will provide step-by-step instructions for each of the examples in the book.

Software/hardware covered in the book	Operating system requirements
Microsoft Sentinel	No operating system requirements
Azure Logic App	No operating system requirements
Azure Log Analytics	No operating system requirements
Azure Active Directory	No operating system requirements
PowerShell	Windows, Linux, MacOS

If you are using the digital version of this book, we advise you to type the code yourself or access the code from the book's GitHub repository (a link is available in the next section). Doing so will help you avoid any potential errors related to the copying and pasting of code.

Download the example code files

You can download the example code files for this book from GitHub at `https://github.com/PacktPublishing/Security-Orchestration-Automation-and-Response-for-Security-Analysts`. If there's an update to the code, it will be updated in the GitHub repository.

We also have other code bundles from our rich catalog of books and videos available at `https://github.com/PacktPublishing/`. Check them out!

Download the color images

We also provide a PDF file that has color images of the screenshots and diagrams used in this book. You can download it here: `https://packt.link/Uz9ge`.

Conventions used

There are a number of text conventions used throughout this book.

`Code in text`: Indicates code words in text, database table names, folder names, filenames, file extensions, pathnames, dummy URLs, user input, and Twitter handles. Here is an example: "Instead of the `items('For_each')?['additionalData']?['MdatpDeviceId']` expression, we can see `additionalData.MdatpDeviceId` as dynamic content."

A block of code is set as follows:

```
split('Incident 1;Incident 2;Incident 3', ';') - result will be
join(body('Entities_-_Get_Hosts')?['Hosts']?[' friendlyName'], ', ')
["Incident 1", "Incident 2", "Incident 3"]
```

Any command-line input or output is written as follows:

```
$GraphAppId = "00000003-0000-0000-c000-000000000000"
$PermissionName1 = "User.Read.All"
$PermissionName2 = "User.ReadWrite.All"
```

Bold: Indicates a new term, an important word, or words that you see onscreen. For instance, words in menus or dialog boxes appear in **bold**. Here is an example: "We utilize the **Select** and **Create HTML table** actions if we want to create an HTML table based on array values, such as a list of all entities and what kinds of entities they are."

> **Tips or important notes**
> Appear like this.

Get in touch

Feedback from our readers is always welcome.

General feedback: If you have questions about any aspect of this book, email us at `customercare@packtpub.com` and mention the book title in the subject of your message.

Errata: Although we have taken every care to ensure the accuracy of our content, mistakes do happen. If you have found a mistake in this book, we would be grateful if you would report this to us. Please visit `www.packtpub.com/support/errata` and fill in the form.

Piracy: If you come across any illegal copies of our works in any form on the internet, we would be grateful if you would provide us with the location address or website name. Please contact us at `copyright@packt.com` with a link to the material.

If you are interested in becoming an author: If there is a topic that you have expertise in and you are interested in either writing or contributing to a book, please visit `authors.packtpub.com`.

Share your thoughts

Once you've read *Security Orchestration, Automation and Response for Security Analysts*, we'd love to hear your thoughts! Scan the QR code below to go straight to the Amazon review page for this book and share your feedback.

https://packt.link/r/1803242914

Your review is important to us and the tech community and will help us make sure we're delivering excellent quality content.

Download a free PDF copy of this book

Thanks for purchasing this book!

Do you like to read on the go but are unable to carry your print books everywhere? Is your eBook purchase not compatible with the device of your choice?

Don't worry, now with every Packt book you get a DRM-free PDF version of that book at no cost.

Read anywhere, any place, on any device. Search, copy, and paste code from your favorite technical books directly into your application.

The perks don't stop there, you can get exclusive access to discounts, newsletters, and great free content in your inbox daily

Follow these simple steps to get the benefits:

1. Scan the QR code or visit the link below

https://packt.link/free-ebook/9781803242910

2. Submit your proof of purchase
3. That's it! We'll send your free PDF and other benefits to your email directly

Part 1:
Intro to SOAR and Its Elements

In the first part, we will introduce cybersecurity and explain why we are speaking about new security tools such as SOAR, as well as introducing SOAR and its importance. In addition to this, we will go over the main elements of SOAR and why they play such a crucial role in SOCs.

This part contains the following chapters:

- *Chapter 1, The Current State of Cybersecurity and the Role of SOAR*
- *Chapter 2, A Deep Dive into Incident Management and Investigation*
- *Chapter 3, A Deep Dive into Automation and Reporting*

1
The Current State of Cybersecurity and the Role of SOAR

Ransomware, data leaks, phishing, denial of service… these are some of the terms that everyone, even those who aren't in the IT industry, will have repeatedly heard in the last few years. Everyone has received an email from a Nigerian prince or some long-lost rich, relative from Africa at least once. These are basic examples of cyberattacks called phishing attacks, which still have an acceptable success rate. If we were to talk about more tailored phishing attacks (common ones being a request to change your password or a notification that your account will be deleted if you don't click on a link), those have an even better success rate – why is that so? Because bad actors are smart.

The first aspect to consider is that they will use many techniques to make their email seem as legitimate as possible, and the second, which is not connected to IT, is the psychological part. The psychological part manifests itself in a few different ways. It can be someone pretending to be your boss (using spoofing methods), an email containing a sense of urgency, or an email sent at the end of working hours when employee concentration is at its lowest. Because of this, organizations are on the lookout for more advanced systems to help them respond to these in a matter of minutes. That is where **Security Orchestration, Automation, and Response (SOAR)** comes in to save the day.

In this chapter, we will cover the main aspects of changes within cybersecurity and how those changes impact our everyday lives. A few years back, cyberattacks mainly impacted organizations, but today, their impact is felt by ordinary people as well. And this is something that will not change overnight. As one way of fighting back and improving their security posture, organizations can use many security tools. One of them is SOAR, and we will explain why SOAR is a must in every organization today.

In a nutshell, this chapter will cover the following main topics:

- Traditional versus modern security
- The state of cybersecurity
- What is SOAR?

Traditional versus modern security

Security plays a significant role in our everyday lives. Even from the start of civilization, security played a role in that people built their fortifications. If we go back through history, we can see how people built their fortifications on the top of a hill or on a river fork, or if something of this kind was not applicable, people dug canals around fortifications, built big walls, and so on. All this had one thing in common – the aim of securing the people and their properties against attacks from other tribes or countries.

As those fortifications were built, attackers always sought a way to penetrate those defenses. Some of them were massive attacks directly made on fortifications, sending a single person to breach the front or back entrance or create a diversion.

Probably the most famous of these, with the equivalent in IT appearing every day, is when ancient Greece attacked Troy. Because of Troy's fortifications, Greece couldn't penetrate the city, even though they had a massive army and the numbers were on their side. That all changed when Odysseus came upon the idea of a diversion. Greek forces pretended to retreat and left a giant wooden horse as a present from the gods to the people of Troy. And what did they do? The people of Troy took that wooden horse into the city. They didn't know that Odysseus and his best fighters were hiding inside that wooden horse. In the early morning, while everyone was sleeping, Odysseus and his selected army exited the wooden horse and opened the door for the rest of the army to enter Troy. After that, all the defense mechanisms in place fell apart, and Troy was defeated.

If you are in cybersecurity, even if you don't know this story about Troy, you will be aware of what a Trojan horse is: a term for malware that misleads users about its true purpose. While it appears to be secure software, it can contain malicious code. It works in much the same way as it did 3,000 years ago.

We can see that many types of historical attacks and defenses are similar throughout history; the only part that changes is how they are performed. We can look at a full army attack on a fortress as a **Distributed Denial-of-Service** (**DDoS**) attack, a Trojan horse as a payload being delivered, a ransomware attack as Vikings asking for gold and valuables to halt their attack on Britain, a spyware intrusion as sending a spy to gather information on fortress defenses from the inside, and so on. From a defense perspective, we can see how everyone started with a perimeter defense by building walls or creating a fortress at the top of a hill. Then, they moved to layered defense by adding water canals in front of walls. The best example of a historic, layered defense was Constantinople. It started with a single wall, and in the end, it contained a moat, a low wall, an outer wall, and an inner wall. If we look at cybersecurity, we can see that there was a similar approach with a single barrier to protect the perimeter – a firewall. This was followed by adding additional layers such as DDoS protection, a **Web Application Firewall (WAF)**, antivirus solutions, and so on.

Looking at this parallel, we all can agree that these defense strategies weren't enough and that even the most robust defenses fell under heavy attack. Even the great Constantinople, probably the city with the best defenses of all time, fell under heavy Ottoman attacks.

Why? As methods of attack evolved faster than methods of defense, it was harder to cover this gap.

The same is true for cybersecurity. As mentioned, we start with perimeter defense and then add layered defense, but even that isn't sufficient. Methods of attack evolve, and bad actors always find a way to surpass existing systems. One thing is certain: traditional systems are outdated, and many organizations are in the process of updating their cybersecurity as a result.

There are a few reasons why this is happening:

- An important aspect is that people are more aware of how they use their personal information, how it is handled, and how it can be misused. People used to trust websites to use their info internally, but those websites sold that info to advertisement companies. People now expect more rigorous privacy and security for the data they share on websites.

- Second up on the list is reputation. Many organizations that suffer an attack experience a loss of reputation, and in the end, smaller organizations often don't survive these kinds of attacks. The loss of existing clients and the absence of new ones to replace them affect many small and medium organizations after a cyberattack. Big organizations survive more quickly because of their size, but they suffer heavy losses.

- The third is bankruptcy, which is directly connected to ransomware in most cases. First, you need to pay to decrypt your data, and on top of that, you have the cost of not running your business. Coupled with a loss of clients, this will all bring small and medium organizations to their knees very quickly. In addition, these companies that have suffered a successful cyberattack end up having their information shared on the dark web. Consequently, they are often targeted by even more bad actors with financial gain as their motive.

Today, organizations either need to update their defense strategies to stay ahead of bad actors or risk a significant cybersecurity incident resulting in considerable financial losses – initially or in the long run.

The current state of cybersecurity

The last few years have changed how businesses operate, and standard working will never be the same. Digital transformation and the COVID-19 pandemic have foundationally changed the way that we work. Modern tools for collaboration, such as Microsoft Teams, Slack, Zoom, and so on, make it possible for people to work from any location and still relate to their peers. When the COVID-19 pandemic started, everyone had to work from home. And something that started as a temporary solution has changed how people work permanently. However, it hasn't just changed the way people are working.

It has also changed how people connect and what network they use – it has changed cybersecurity. A traditional perimeter does not help anymore; people are expected to be outside their *bubbles*, and we must find new ways to protect them. The second thing to consider is that people don't just use corporate devices to connect to corporate resources: they use personal devices as well.

Creating boundaries is becoming harder and harder, and organizations must find a new way to protect their resources. Traditional systems aren't enough anymore. The first tools that people are turning to have been available for years in the market, such as **Mobile Device Management/Mobile Application Management (MDM/MAM)**, **Multi-Factor Authentication (MFA)**, **Endpoint Detection and Response (EDR)** platforms, **Data Loss Prevention (DLP)**, and so on.

Introducing more security tools and hardening the working environment has a direct impact on productivity. Employees are expected to enroll devices to MDM, set up and pass MFA, avoid copying data to USBs, refrain from continuing their work on other devices, and avoid sharing any links with anyone. This significantly hampers the ability of employees to collaborate efficiently. Cybersecurity experts need to find a golden middle ground between productivity and security; often, this equates to sacrificing security under this pressure until a cybersecurity incident occurs.

To be able to detect security incidents as they happen, more advanced solutions are required: traditional ones such as **Security Information and Event Management (SIEM)**, more modern ones such as **Extended Detection and Response (XDR)**, and the Zero Trust methodology. SIEM allows us to collect logs from various solutions and correlate these events to detect threats more easily. However, on its own, it is ineffective. SIEM tools are only as good as the events they have as logs. We also need to have excellent **Security Operations Center (SOC)** analysts who can define detection rules, do cyber threat hunting, and react to security incidents in these SIEM solutions. This is why most new SIEM solutions add **Artificial Intelligence (AI)**, **Machine Learning (ML)**, **User and Entity Behavior Analytics (UEBA)**, **Threat Intelligence (TI)**, and so on, into the mix to help with detection – but what about the response? How do we acknowledge and resolve security incidents?

One of the more modern tools is XDR – this is not a single tool but a group of tools that work together to correlate cyber threat detections. In most cases, XDR will cover identities, emails, endpoints, servers, and cloud workloads. It will use AI and ML in the background to connect security incidents from these layers, which are often handled separately by different solutions, into a single incident that outlines the kill chain of an attack as it happens throughout an organization. While XDR is a must-have solution for most organizations, it still doesn't cover the whole stack of security. You cannot ingest TI data, firewall logs, third-party solution detections, and so on. Typically, XDR will be connected to SIEM for correlation with other sources.

One thing we have seen with XDR is a change in the complexity of organizations' cybersecurity. 10 years ago, organizations did not use the same vendor for different layers of protection. The idea was that if one failed, you would still have another vendor in line for protection – but how wrong was that?

First, our security experts had to learn to work with and manage multiple solutions and vendors. Multiple portals would therefore need to be logged in to daily. For big organizations, the number of security solutions and vendors used could exceed 40! And second, those solutions did not *speak* to each other. That means that they did not share intelligence; they did not correlate their shared data. Without SIEM collecting events from all devices, it was almost impossible to make connections between different security incidents. XDR changed this, as the idea behind it has been for solutions to speak with each other, share intelligence, and correlate events for better detection. Another significant benefit is that it is all in one portal, which is essential for security experts to focus on one unified product and not on five different ones.

Why is it essential to find new ways to protect organizations? Because bad actors are improving their game daily. Just in the last few years, we have had significant cyberattacks, including the Colonial Pipeline ransomware attack, the Maersk ransomware attack, the SolarWinds breach, and the Log4j vulnerability, plus many data breaches in which bad actors have stolen terabytes of personal data. These are only some of the attacks that have been top news worldwide. Even people who don't know what a cyberattack is have started asking questions about what is happening. The reason for this is the significant impact of each attack. The Colonial Pipeline attack raised a lot of concern and panic among people in the United States. Because of this attack, a few states even reported shortages of fuel. Even though Colonial Pipeline paid the ransom (in total, around 5 million US dollars), restoring operations took them a few days. As a direct connection to the attack, fuel prices in most of the United States went up.

This is only one of the examples of how a cyberattack on critical infrastructure can impact an organization and a whole country. Let's consider that most of the critical infrastructure in countries (electricity, water, fuel, gas, etc.) is controlled using computers. We can see why staying at least one step ahead of bad actors is crucial.

There are many different figures for the average cost of a cyberattack, and in most cases, the average cost is around $4 million. This cost is not only connected to paying a ransom but also returning to an operational state, plus the cost of losing customers. If we take a look at the Marriott hotel data breach, the total cost at the end could be in the billions, as we include the GDPR and user lawsuits. We can say that, on average, we have millions of reasons to think about cybersecurity at a time.

However, cyberattacks don't just impact organizations; they are methods of modern warfare. We have had a few examples throughout history, but the latest one is probably the best example. As the Ukrainian-Russian war started, it didn't start solely with typical military conflicts – guns, tanks, planes, and so on. Cyber warfare was a big part of it, and numerous attacks on Ukrainian infrastructure were reported.

Considering that we have more and more drones in the sky that are remotely managed, it shows us how serious it can be in the future if technological infrastructure is not protected.

While we can invest a lot of money into security equipment, we still have two significant issues at the top of the list regarding *how a cyberattack starts*. The first will be misconfiguration, and the second will be the user.

As mentioned, many organizations invest a lot in security equipment, but not in security experts or their personnel so that they can learn how to configure solutions correctly. Even a minor misconfiguration can affect the system in a manner that will leave a backdoor that a bad actor can use. Hiring security experts and continuous investment in cybersecurity personnel is more important than security solutions. Cybersecurity personnel must stay ahead of bad actors to protect critical infrastructure. While AI and ML play a significant role in cybersecurity, they will (maybe) never be able to replace security experts. Most sophisticated attacks are not initially detected by cybersecurity tools but rather by experts hunting for anomalies in raw system logs.

Users are probably the most considerable cybersecurity risk each organization faces. It is a common saying in cybersecurity that in each organization, there is at least one user who will click on every link. That is why phishing attacks are still the most common attacks on organizations. Every organization must invest in user education to reduce the risk of users clicking on a link in an obvious phishing email or downloading attachments from unknown sources. It is a long process to educate users and still, the risk will exist. As mentioned earlier, bad actors are smart and target users strategically – for example, when they know their focus will be at the lowest at the end of working hours.

On top of that, think about every conversation had with users – passwords. It is common for users to pick the same password for business and personal use and reuse it across all platforms. Some people use two different passwords, but rarely three or more. This directly impacts an organization's security because many platforms don't have advanced password protection – but that is not the only problem! Users incorporate personal information when creating these passwords (such as a place of birth or residence, names of pets or children, important dates, and so on) and then have all of that information publicly available on social media (pictures, About Me, favorite movies, quotes, and more). Because of all this, it is easy for bad actors to strategize their attacks. First, they have all the necessary info to create a dictionary for brute-force attacks on social media. Second, they can use a less secure platform to perform that attack and reuse the password on corporate logins. This is essentially why many organizations implement MFA.

The biggest challenge for modern SOCs is the high number of raw data and security incidents. This affects the time needed to acknowledge and respond to security incidents. The initial triage of an incident can take some time, even an hour, if a SOC is inefficient or there are not enough SOC analysts (which is more common). This can lead to detecting the cyberattack too late, and the attack can spread through the system.

Would it help if we could automate everyday tasks that our SOC analyst performed as part of the initial triage so that the SOC analyst took over once the initial triage had automatically been done? This is where SOAR comes into play!

What is SOAR?

SOAR is a set of security features that helps organizations collaborate on incident investigation and automate certain actions that SOC analysts perform. As the end goal with SOAR, we want to achieve a faster **mean time to acknowledge (MTTA)** and **mean time to respond (MTTR)**. The MTTA and MTTR are the two most important measurements for a SOC.

The main elements of SOAR are as follows:

- Incident management
- Investigation
- Automation
- Reporting
- TI and **Threat and Vulnerability Management (TVM)**

> **Important note**
>
> We will touch on reporting as a separate topic in *Chapter 3*. We will also discuss TI and TVM through automation in *Chapter 6*.

SOAR is so important due to the increasing number of events to analyze and security incidents to investigate, and the deficit of security experts to perform the job. If you look at SOAR as a complete replacement for SOC analysts, you couldn't be more wrong. SOAR is probably a SOC analyst's best friend and provides the SOC team with the ability to analyze threats faster. SOAR as a tool and SOC teams can reduce the MTTA to a few minutes and the MTTR from hours to minutes!

How? The main aspect of SOAR is action automatization. That means that any task that the SOC team repeatedly performs during an incident should be automated. First, this will save time for SOC analysts – plus, we don't need to worry about whether SOC analysts may forget to perform any tasks. Second, we can carry out the initial triage, and based on the input, we can auto-close false positives so that the SOC team doesn't even need to work on them. Third, once the incident is assigned to SOC analysts, they can automatically see the enrichments made by automation to that incident. This will empower them to analyze and react to incidents much faster.

Incident management is an essential aspect of SOAR as well. If we want our SOC analysts to respond to incidents effectively, they need to have the *space* in which to do so. Not only space but also features will empower SOC analysts. These features include an incident overview, the possibility to increase or decrease the severity rating, close an incident, assign an incident owner, see more details, quickly navigate an investigation, comment on incidents, and so much more.

The reason why an investigation is essential is that the SOC team needs to gather as much information as early on as possible for an effective response. That can be through looking at similar incidents; checking what accounts, hosts, and IPs were included; whether those IPs, hosts, and accounts are known or not; how they connect with other data in the solution; and the ability to perform threat hunting. In addition, reporting, TI, and TVM provide even more insights to the SOC team to help perform an incident triage quickly and correctly.

So… do I need solutions such as XDR, SIEM, and so on? Or is SOAR enough?!

The quick answer is yes! These technologies differ in how they handle one common task – quickly and efficiently protecting your organization against threats.

Let's look at the current situation in the market. We will see that many SIEM vendors either developed their own SOAR solution or bought a SOAR solution and integrated it into their environment. Microsoft Sentinel uses the power of Azure and Logic Apps for automation. Palo Alto bought Demisto (now called Cortex XSOAR) and integrated it into their XDR offering. Splunk bought Phantom and integrated it into their SIEM offering (now called Splunk SOAR). IBM bought Resilient and merged it into their SIEM offering (now called IBM Security QRadar SOAR). And the latest example is Google's acquisition of Simplify and how they have merged it into their offering.

In all these examples, we can see a few trends. The most important one is that the future is to merge security tools into one so that you can manage your security completely in one place. The boundaries between security tools are receding constantly, and tools such as XDR, SIEM, SOAR, and so on are integrated more and more to provide a native, one-portal experience to organizations. The well-known line from Lord of the Rings is "*one ring to rule them all*," and in security, it will be "*one tool to rule them all*."

OK, so SOAR is here to stay – but what are the typical use cases?

- **Incident enrichment**: Here, we will use the information found using TI and TVM solutions to enrich incidents with more data:

  A. Is that hash or IP malicious? Check using TI and, based on this, you can escalate the incident or even close it if all the data is well-known to your organization.

  B. Does that host have any vulnerabilities? Check using TVM whether any **Common Vulnerability and Exploit** (**CVE**) is connected to the host and decide how to proceed.

  Here, we can see how we can use automation to quickly grab that info on incident creation, and when the SOC analyst picks up that ticket, the data will be there. As a result, the SOC analyst doesn't need to perform an initial triage, thus saving time. Based on this info from automation, we can make faster decisions on how to proceed with an incident.

- **Incident remediation**: Let's say that, from the first step, we find out that an IP is malicious or that a host has a critical CVE. As a response, we can run automation that will block that IP in our firewall or EDR solution, or we can isolate that host so that it cannot cause any damage. This is done from the same portal; there is no need to go to different solutions, copy the IP, and then block it. With a click of the playbook, all will be done.

- **Reduce fatigue by reducing the number of false positives**: SOC teams have significant issues when solving false positives. It takes time to open each incident, check whether it is connected to our known data, and close it. What if the SOC analyst didn't even need to look at it? Automation can be run to check for well-known data. If it is connected to well-known data, we can auto-close an incident: this means zero engagement from the SOC analyst.

The examples mentioned are clear examples of how tools such as SOAR can help improve the MTTA and MTTR. Instead of repeating tasks, the SOC can focus on high-severity and true-positive incidents. It's a well-known fact that good SOC analysts will burn out after a few years, and organizations will need to bring in new analysts who need to be onboarded and taught the SOC's tricks. SOAR will help to decrease pressure on the SOC by reducing fatigue. With it, mental health improves, and SOC analysts don't burn out. That also means they can perform their job longer, be more satisfied, and focus on the tasks ahead. By reducing the number of events and incidents that a SOC analyst needs to resolve, they can also invest more time into learning about new defense methods. Overall, the losers in this picture are the ones who *should* be losing out – the bad actors.

Summary

This chapter covered the importance of improving your security strategy and keeping your organization's security one step ahead of bad actors. We saw how the traditional method of protection is outdated, a perfect scenario for bad actors, and how they can utilize even the most direct attacks to take down organizations.

Throughout the chapter, we also touched base on the state of cybersecurity, how organizations are changing their strategies, and how new tools such as XDR are emerging. Equally, these new tools directly influence SOC teams being overloaded because more tools equal more events, which equals more security incidents. Since there is a significant gap in the market for security experts – and it takes a long time to investigate the share volume of events and incidents manually – there is a need for *help*.

This is where SOAR jumps in and helps organizations automate everyday tasks. This directly impacts the efficiency of SOC teams, reducing the MTTA and MTTR and overall SOC fatigue. We then introduced simple use cases for SOAR, such as incident enrichment, remediation, or auto-closure. Later in the book, we will use similar cases to go through how to set up automation step by step.

The next chapter will go through some of the most well-known SOAR tools. These solutions are often part of more comprehensive SIEM tools, and we will explain how those SIEM tools were nudged forward as the *ruling* security solutions. We will go through the main aspects of SOAR, such as incident management, investigation, and automation, and how these features are utilized in the day-to-day activities of SOC teams.

2

A Deep Dive into Incident Management and Investigation

The previous chapter touched on the importance of cybersecurity today and why investing in modern security tools and education is essential. We also covered the main pain points of SOCs today, such as the never-ending growth of security incidents and security-related logs, the number of tools, and staying arm-to-arm with bad actors.

As established, the SOAR solution consists of a few different elements, such as incident management, investigation, automation, reporting, **threat intelligence** (TI), and **threat and vulnerability management** (TVM). Each of these plays an important part in a SOC and makes SOCs more efficient and effective. This is critical today since SOCs are overloaded with security incidents and signals that they need to handle.

In this chapter, we will focus on incident management and investigation and why it is an important segment when we think about SOAR. We will also touch base with some important aspects to understand when working with an incident, such as the **National Institute of Standards and Technology** (NIST) and **SysAdmin, Audit, Network, and Security** (SANS) frameworks. The topics that will be covered in this chapter are as follows:

- Understanding SOC tiers
- Understanding incident management
- Investigating NIST and SANS incident management frameworks
- Incident tasks – to do or not to do
- The incident queue as the investigation starting point
- The incident investigation process
- Threat hunting – the next step

> **Reminder**
> *Chapters 2 and 3* will be a theoretical overview of what SOAR elements are while *Chapters 4* through *9* will focus more on discussing these elements with in-product and hands-on examples, as well as expert tips and tricks.

What are SOC tiers?

As we will mention the terms SOC, SOC analyst, and others many times in the book, it is best to explain the main SOC tiers and their responsibilities. These tiers are positioned to help SOCs divide tasks among SOC analysts:

- **Tier 1**: This is the first tier and it is primarily responsible for triage incidents as they are happening. They will collect as much information as they can about the incident so that they can plan whether that incident should be investigated further. This is an entry-level role in SOCs for those with the least experience.

- **Tier 2**: The second tier *kicks in* when Tier 1 decides that a specific incident needs to be investigated further. One of the responsibilities is coordinating responses with different teams in the organization. As these incidents are more likely to be genuinely positive, this tier requires a more experienced SOC analyst.

- **Tier 3**: This tier is dedicated to experts in the field. Their primary responsibility is to perform threat hunting. Tier 3 goes through *a sea of logs* trying to find possible IoCs and **Tactics, Techniques, and Procedures** (**TTPs**) and works with other tiers so that they can respond to possible threats as soon as possible.

Next, we will look at incident management and why it is an important aspect of SOAR.

Understanding incident management

Incident management is the process we go through from incident detection to the time the incident is resolved. In SOC, this is where SOC analysts will be able to monitor incidents as they are created, filter incidents based on conditions, search through the incidents, and perform fast incident actions.

Without incident management, our SOC analysts wouldn't be able to see what incidents were created and from where they would need to start the incident investigation.

The primary purpose of incident management is to do the following:

- Detect the incident

- Investigate the incident

- Contain and recover from the incident

- Document the incident

Why do we need incident management in SOAR?

The main purpose of SOAR is to make the daily tasks of a SOC easier to handle. Let's see how SOAR accomplishes this purpose.

Imagine that the SOC team doesn't have an incident management solution. An incident is detected, and the SOC analyst takes it over for investigation. Because there is no incident management, no one is aware that the incident is already being analyzed and that more than one SOC analyst can investigate the same incident. We don't know at what stage this incident investigation is, nor do we have a central point to see all incidents and filter them to see what is assigned to each SOC analyst. This means that the SOC will need to have some system (in most cases, an Excel table) that consists of a list of incidents and who is analyzing them. SOC analysts need to keep updating that list (I think that I was clear earlier on how much IT professionals adore admin stuff).

The additional issue here is that SOC analysts will lose a considerable amount of time looking for data that should be available at first glance. It will also involve multiple clicks to get that information.

This is not the best practice in cybersecurity because the SOC analyst needs to have an organized approach to handling incidents, prioritizing them, and improving **Mean Time To Acknowledge (MTTA)** and **Mean Time To Resolve (MTTR)**. MTTA is the time it takes for the SOC to acknowledge that an incident needs to be investigated. This is important as it can tell us how long the incident was active in our system before the SOC became aware of it and began working on it. A high MTTA, for example, can show us that SOC analysts are overwhelmed with incidents and cannot start investigating on time. The MTTR (also known as the **Mean Time To Close (MTTC)**) will show us how much time it took the SOC to finish the investigation, take appropriate actions, and close it in the system. This is an important metric because it will show the average investigation time, and based on this, you can plan the SOC size. If you have six SOC analysts, and you have 60 incidents a day on average, and the SOC analysts work in three shifts of 8 hours – two analysts in each shift – with a 60-minute average of MTTR, this will show that one shift can cover 16 incidents – that is, 48 incidents in a day, which means that 12 incidents cannot be investigated in a day. And 12 incidents per day is 84 incidents in a week. Any delay in incident response can significantly impact the level of the breach. A few minutes can be the difference in detecting and stopping the breach.

Incident management offers SOCs an organized approach to handling incidents. Based on the data that incident management solutions offer, SOC analysts can quickly identify incidents with higher severity or whether they include high-value assets. This will impact MTTA as SOC analysts can have a more organized approach.

As incident management has the option to see whether the analyst (incident owner) is assigned and the incident's status, it is easy for SOC analysts to know which incidents are being investigated and which ones have been opened with no one working on them. SOC analysts will have a better option to coordinate and not have two SOC analysts working on the same incidents unknowingly. Even better, thanks to automation, it is possible to assign incidents automatically. This makes SOCs even more aligned. We will cover more about automation in *Chapter 3*.

Now, let's imagine that our SOC has an incident management solution. Incidents are assigned to SOC analysts on creation, and the SOC analyst can get notifications about new incident assignments using automation. Once the SOC analyst logs in to SIEM/SOAR, there will be a list of open incidents, and it will be easy to filter those by owner, severity, and more to find specific accounts, tags, and so on. Each incident has a status (*new*, *in progress*, and so on), and it is easy to keep track of SOC metrics (MTTA and MTTR). SOC analysts will use comments to keep track of new developments in the incident investigation so that if the incident needs to be transferred to someone else, the new SOC analyst will be able to read comments and quickly be up to date with the incident status and progress.

Once we are done investigating the incident, we can write the lessons learned and close the incident with the correct classification (true positive, false positive, and so on). This last step plays a vital role in a long-term SOC strategy. With correct classification, we can easily keep track of several false positives that SOC analysts must investigate and make correct plans to lower that number – either by fine-tuning detections or by automating the response.

Exploring incident management features

As explained in the previous section, if a team is working on a project and doesn't have a list of who is responsible for which segment of the project, it can be expected that two or more members will be working on the same thing without sharing information between each other or without making sure that tasks aren't overlapping. Therefore, in most cases, organizations use tools and project managers to ensure everyone is aware of their tasks and who is responsible for what segment.

The same applies to the SOC team and how they work. Most SOCs have multiple members, and they need a tool to help them organize incident management inside the organization. As we mentioned previously, the number of incidents is rising every day, and with the proper tools, SOC analysts can perform their tasks efficiently. For years, IT departments have been utilizing service desk tools to monitor tickets, and this same approach is used in SOARs to perform incident management. Even some traditional service desk tools such as ServiceNow and Jira can perform incident management with integrations with many SIEM tools.

So, what does the incident management tool contain? Several features are important in incident management:

- Incident queue
- Ownership
- Severity
- Status
- Comments
- Tags
- Tasks

- Investigation space
- Lessons learned

Also, many organizations are following their incident response with well-known frameworks, such as NIST or SANS. For them, it is important to categorize the incident life cycle as per the framework's steps. This can help the SOC team define how to handle the incidents by following an agreed-upon process.

Incident queue

An incident queue normally contains a list of all incidents detected on the system. It must be easy to navigate through the incident queue and get more details about the specific incident we want to focus on. To get more details about each incident, the information should be just one or two clicks away, and it must be easy to return to the previous screen. Complicated interfaces or those with too many details can take more time for SOC analysts to process and for them to find the important information needed to decide what to do with the incident in the initial stage. Usually, SOAR tools will have a dedicated incident investigation queue that will contain more data about a single incident that we will be using in the investigation phase.

Since this queue contains all incidents, we must have the option to filter these incidents per specific scenarios. That can be to filter per a particular owner of the incident, severity, or status. This is important when we want to look at several incidents with high severity assigned to certain SOC analysts. This also can be done using reporting, which we will cover later in this book, but sometimes, we don't want to leave the view we are currently in – we want to be able to do it from the incident queue with as few clicks as possible.

One more important aspect is to have a search option, which should have the option to find a specific incident by incident title or specific identifier (ID), or even to have more advanced options to search by specific elements of the incidents such as IP, account, host, and so on. This is because we want to find more incidents with specific names or accounts so that we can easily and quickly see whether it is a common incident, whether the incident hasn't occurred for a while, whether the account was a part of other incidents, and so on. Some aspects of SIEM/SOAR will have this option, such as similar incidents or account pages, but sometimes, we want to find it fast, through the incident queue, and not via a few clicks to get to that information.

An incident queue's final important aspect is its ability to perform a quick action on the incident. It can be to close it, change the severity or owner, or run automation for automated enrichment or a response. Let's say, for example, that we filtered certain incidents. We searched for incidents that contained specific titles, and we included open and closed incidents. We saw that it is a common incident and, in most cases, a false positive. Then, we filtered by that certain account and saw that it is always a false positive. Maybe this information is enough for us to decide that this incident is also a false positive, and we want to close this incident with as few steps as possible. Since every second is important and we don't want SOC analysts to go through extra steps to close incidents, having the option to do so from the incident queue can make a difference.

Ownership

As a SOC can have multiple SOC analysts, it is important to know who is working on the specific incident. This is done by assigning an owner to the incident. That owner is then responsible for the incident until the incident is either transferred to the new owner (for example, from Tier 1 to Tier 2) or the incident is resolved.

Since one SOC analyst can have multiple incidents assigned to its name, we must have the option to filter only incidents for that specific SOC analyst. Some SOARs have an additional overview page that will show only open incidents from that SOC analyst, with no need to filter them.

One thing that is becoming more and more popular is automating the assignment of incidents as they are created. This helps reduce the risk of having unassigned incidents. More about how to automate the incident assignment will be covered in *Chapter 3*.

Severity

The severity of the incident will give us important information on how serious the incident is. Some standard classifications are **Low**, **Medium**, and **High**, while some vendors add severity classifications such as **Informational** or **Critical**. This helps the SOC analyst prioritize incident investigations. If we imagine that one SOC analyst can have dozens of incidents assigned at the same time, filtering from High to Low in prioritization can help the SOC analyst resolve higher impact severity incidents first.

The severity level must be able to be changed as we may want to decrease or increase the severity if we find important information during the initial triage. For example, let's say we have low-severity incidents. During the initial investigation, we find that the IP address in the incident is connected to recent global attacks. In this case, we want to increase the severity so that we have the correct classification. Correct classification is important initially as we know that certain incidents have higher priority for investigation, as well as later when we are doing documentation and reporting.

Status

Not all incidents have the same status. On creation, incidents normally have a status of *new*. This means that the incident was just created, and the SOC analyst hasn't started working on it. Once SOC analysts take the incident under review, it is not *new* anymore. In this case, many vendors use a status of *in progress* or *active*, which means it is currently under review. The final stage is *closed* or *resolved*, which is marked when the SOC analyst finishes the investigation. An important aspect of the final stage (*closed/resolved*) is to classify that incident as a true positive, false positive, or something else. This is important as more and more vendors are adding AI and ML to help SOCs fine-tune their detections. Also, it is important for reporting purposes so that SOC managers can analyze and determine whether an incident was a true or false positive. A true positive incident is an incident that contains an actual breach and the actions performed are indeed malicious, while a false positive incident is an incident that is detected, but the action performed is expected and not malicious. Some actions performed by employees can be seen as safe, but if they are performed by an attacker, they are malicious. Normally, we would like to correlate multiple signals or actions to be able to differentiate expected user behavior from malicious-attacker behavior. For example, five unsuccessful logins before a successful login could happen because the user had issues remembering their password. But if we had the same situation occurring alongside potentially malicious software such as Mimikatz or PSExec being detected on the computer, it could be hinting at a false positive detection of a brute-force password attack, with malicious software commonly used for lateral movement techniques.

A practical example of why it is important to have correct statuses is reporting. The first example is how we can track the MTTA and MTTR. The MTTA will be when an incident is acknowledged and moved to an *active* stive, while the MTTR will be when an incident is *closed*. With this, we can keep track of these important metrics in the SOC and ensure we don't have a long MTTA or MTTR. A long MTTA will indicate that we maybe have too many incidents assigned to our SOC analysts and that they cannot manage them sufficiently to have a shorter MTTA. It is usually an indication that we need more SOC analysts, we need to tune our detection logic, or we are underutilizing automation for incident triage.

Another example is to monitor how many incidents are true positives versus false positives. If we see that we have too many incidents closed as false positives, we probably need to revisit our rules written to detect events that will create incidents. We may need to fine-tune those rules or turn them into queries that we will use to hunt for suspicious activities. If we still want to have records of these false positives for future investigations and hunting, we can utilize automation to resolve known false positive incidents.

Comment

Comments on incidents allow SOC analysts to leave important data on the incident. They are also used with automation to give automation verdicts, to log what was performed on the incident, whether it is something known such as a bad actor, and so on. Many SOCs have a few tiers that work on incidents and comment on what was performed and what the result was. This helps the next tier level quickly understand why the incident escalated further.

Today, we use automation on many incidents to do the initial triage of the incident or to enrich the incident with further data. Automation mostly uses comments as an area where it will write down what is performed on the incident. This is an important aspect of SOAR, and we will cover it in more detail in *Chapter 3*.

Tags

In addition to adding comments, it is also popular to add tags to the incident. A tag can contain specific information about the incident or an option to group and identify similar incidents. It can also add important information, such as whether the host is part of **high-valued assets** (**HVAs**) in our organization by adding an HVA tag, or whether the user is a **very important person** (**VIP**) by adding a VIP tag.

Tasks

Each SOC analyst handles numerous different incidents. Those incidents will have some common steps that the SOC analyst performs on each new incident, but then some steps are performed only on certain incidents. If we expect SOC analysts to remember all those steps in a matter of minutes, the SOC has even more significant problems than we realize.

Therefore, tasks are one of the crucial elements of incident investigation. Instead of trying to make SOC analysts remember each step from each incident, you can assign steps that need to be performed on the incident. In this case, once the SOC analyst takes over the incident, a list of steps will be assigned to the incident, and the SOC analyst can start triaging one step at a time.

These tasks need to have some important functionalities; a list is good but not good enough. Tasks need features such as marking steps as *done* or *in progress*, making certain tasks mandatory (cannot close/resolve incidents until mandatory steps are done), adding comments to the step, and so on. This can help us understand whether certain steps are unnecessary, whether we need to perform them sooner or later, or whether they can be automated to save valuable time in the incident investigation process.

Since tasks can be one of the crucial elements in the SOC, we will cover tasks in more detail later in this chapter.

Investigation

When working with incidents, SOC analysts must have the option to investigate the incident properly. This starts with a more detailed overview page, as well as options to track notable details such as IP, host, or account. Later in this chapter, we will cover investigations as a part of the system, how to perform them, and their main aspects.

Usability

When designing systems, it is always important to be able to get to the desired location in the fewest possible clicks. This is a crucial aspect of incident management since a timely response to the incident is mandatory, and even seconds can make a difference between remediating the threat before it has an impact and causes a big breach.

The **user interface** (**UI**) must be clean and contain only the necessary information. Any unnecessary information presented can take the SOC analyst's attention away, which is something we don't want. Initial information must be basic, with the option to see more detailed information. An example could be an entity that's been detected (host, account, or IP). In the initial overview, we want to see what entity it is, and with a single click, it will lead us to a specific entity page that will show us more information about that entity.

It must be easy to navigate between the incident queue and the incident investigation page. Suppose the UI is too crowded and does not follow UI methodology throughout the whole system. In that case, it can be problematic for SOC analysts to remember where specifics are, especially for new SOC analysts.

In the next section, we will focus more on incident management frameworks – in this case, NIST and SANS – and how they can help SOCs build incident management, from incident detection to incident documentation.

Investigating NIST and SANS incident management frameworks

As we mentioned at the beginning of this chapter, there are a few incident management frameworks that organizations follow through the incident life cycle. In this chapter, we will introduce the NIST and SANS frameworks.

The NIST framework consists of four steps, while the SANS framework consists of six steps. In the following table, we can see that both frameworks are similar and follow the same principle:

NIST Framework	SANS Framework
Step 1 – Preparation	Step 1 – Preparation
Step 2 – Detection and Analysis	Step 2 – Identification
Step 3 – Containment, Eradication, and Recovery	Step 3 – Containment
	Step 4 – Eradication
	Step 5 – Recovery
Step 4 – Post-Incident Activity	Step 6 – Lessons Learned

Table 2.1 – Comparing the NIST and SANS incident management frameworks

Which framework should we choose? Or should we, as the SOC, create our own framework?

NIST and SANS both provide crucial steps in incident management and if you choose either of them, you will not be wrong to do so. It is important to speak internally about the steps and how you normally go through the incident life cycle and choose a framework that best suits you.

Many organizations use the NIST and SANS frameworks as guidance to make their incident management framework. This will require more time as it needs to be correctly aligned, but some organizations will add more steps, or they will say that five steps are enough since they want to merge eradication and recovery under the same step.

Let's get into these steps!

Preparation

There are many examples where SOCs are either not performing *Step 1* or going through it quickly. In both cases, this can be a big issue later as this step is documenting what our incident management and response would look like.

From an IT perspective, the first and last steps in both frameworks are boring admin stuff, but they are as equally important as all the other steps.

In *Step 1 – Preparation*, we are making documentation of how our incident management and response will look. What does this mean?

We need to have an incident response plan in place that will be followed. We need to know who the important stakeholders that we may need to contact are (common examples are networking engineers and system administrators), as well as those higher in the food chain, such as the **Chief Information Security Officer** (**CISO**). With laws such as GDPR, this is important as companies must report the breach to authorities in time; otherwise, they will be penalized.

We also need to document HVAs and VIPs in the organization. A breach on the financial server differs from a breach on the reception desk laptop. While both can be severe, the financial server will have a more significant impact on the organization.

Another important aspect is that the whole SOC follows the same methodology and uses tags, comments, statuses, and so on in a unified way. If not, this can be a disaster for the SOC. When there is no clear guidance, SOC analysts will use their version of tags and statuses to mark incidents and comments. When transferring incidents between Tiers 1 and 2, those tags, statuses, or comments can mean something different to that analyst. This will take valuable time away from the analyst.

As SOCs are not usually made with only one SOC analyst, they must have a way to communicate and share data easily. Something that happens in the morning can be valuable information for later in the day, and if it's not shared, it can take a longer period to react, or the incident could be ignored.

Detection and analysis/identification

In this second step, we are triaging the incident. Once the incident has been detected, we want to find out more information and enrich that incident. We want to know whether it is a true positive or a false positive. If it is a true positive, we need to perform *Step 3* of NIST or *Steps 3, 4,* and *5* of SANS. If it is a false positive, we need to close it with the correct classification and perform the last step – documenting the lessons learned.

Triaging can be seen as an entry-level SOC job, as junior SOC analysts mostly do it, but the methodology to perform initial triaging must be clear. SOC analysts must follow the same triaging best practices so that if we have a true positive incident and it is escalated to Tier 2, all steps necessary are done correctly. In this step, the use of tasks can be beneficial; we will cover more about tasks later in this chapter.

Suppose we join this step with the next step. In that case, we can put too much pressure on junior SOC analysts to perform containment, eradication, and recovery, which are critical tasks connected to true positive incidents. These tasks must be assigned to more experienced SOC analysts, which is why we have different tiers in the SOC.

Containment, eradication, and recovery

This is where the biggest difference between NIST and SANS can be found. SANS sees it as separate steps (*3*, *4*, and *5*), while NIST puts them all into one step (*Step 3*). This step is performed if we found that the incident is a true positive after performing *Step 2*.

Let's define each of these stages:

- **Containment**: Contains the incident. If the endpoint is breached, isolate the device so that it is not spread across the system – contain the threat. This is one of the steps that we have to perform as soon as possible as we have to stop the threat from spreading through the organization. The best examples of this are worms and ransomware, which can paralyze the whole organization in a matter of minutes if they're not contained.

- **Eradication**: Removes the threat from the system. If the endpoint is breached with malware, remove the malware from the endpoint. Once we contain the threat we are fighting, we have to eradicate it from the system. We have to be sure that it is *flushed* out and that we can start returning to normal.

- **Recovery**: Returns to *business as usual*. In our example, this means to un-isolate the endpoint. After we have contained the threat and we are sure that the threat has been removed from our system, we can start the process of returning to a normal state. In this step, it is also important to continue monitoring to be certain that everything connected to the threat has been removed.

I would agree more with the approach SANS takes here of separating these phases into different steps. Each of these steps is equally important and should be performed in this specific order – first, you contain the threat, then, you eradicate it, and after eradication, you perform the recovery. If you perform eradication before containment, the threat can spread to other systems. Or if we do recovery before eradication, the threat can still be present, even after the recovery. If we take the example of ransomware that is commonly spreading fast through the organization, containing the infected devices is mandatory as the first step; then, you eradicate the ransomware from those systems before going through the process of recovery.

We will talk about *Steps 2* and *3/3, 4*, and *5* in more detail in the next part of this chapter when we cover incident investigation.

Post-incident activity/lessons learned

As we mentioned already in *Step 1* (preparation), many IT professionals see this step as *boring admin stuff* and will typically *fight* not to perform this step. SOC analysts want to quickly jump to another incident as they usually have big queues of incidents to investigate.

But this is one of the more important steps, especially as people tend to look at it as *Argh, why do we need to do it?*

But why is it important? Isn't it more important to triage incidents correctly and perform containment, eradication, and recovery in the right way? Incident documentation will not impact my incident investigation.

If I had one dollar (or euro, Bosnian mark, or gold brick) for every time I heard this....

In reality, this step can drastically improve the organization's MTTA and MTTR. Initially, maybe not, as we will need time to document the incident properly, but in the long run, it can make big differences in the organization. How?

Let's imagine that we are handling the SOC. Usually, we have a few dozen incidents that are raised every day. Some of them are common, and we know how to solve them fast. However, some incidents are either new or don't appear in our incident list that often. If we are utilizing incident tasks, it can help us to refresh our memory on how to investigate those rare incidents. We will cover more about the importance of incident tasks in the next section. But those incident tasks will not always show the whole story. If we had that incident before, we can find it and go through the documented lessons learned from the last investigation and quickly figure out how to approach the incident investigation, where to pay attention, what mistakes we have done to learn from them, and so on. Instead of SOC analysts losing valuable time rediscovering the wheel, it is possible to easily access that information by utilizing lessons learned.

A second important aspect of lessons learned, connected to incident tasks, is that SOC managers can utilize those lessons learned to discover whether incident tasks are assigned correctly and whether there is a need to add additional tasks or maybe delete some tasks. SOC managers can also utilize this for SOC training to show how to perform specific actions based on lessons learned.

When we spoke about the detection and analysis/identification step in the incident management framework, I mentioned that a possible solution could be tasks. But tasks are often misused and can potentially be an issue in SOCs. This is why we will focus on the topic of tasks in the next section and bring a bit more perspective about proper use cases.

Incident tasks – to do or not to do

Recently, I was involved in a project around incident tasks (some know it as incident workflows or incident playbooks) and had the opportunity to speak with many people about it. This is the reason why I decided to have a dedicated section about one important segment in incident management.

As I wrote in the previous section, a few dozen incidents happen daily in our SOCs. Some of them are new, but some of them are only seen sometimes. If there is a new incident that our SOC analyst has never worked on before, it will take some time to investigate it properly. But if someone else had already been investigating that incident before and had written the lessons learned, it would be easy for the SOC manager to create a list of steps that should be taken to investigate this incident. In this case, our SOC analyst will save valuable time figuring out steps that they need to perform, and it is possible to go through the list of tasks that have been assigned quickly. This can lead to drastically lowering the MTTA and MTTR in our SOC.

This is important for the new SOC analysts that we are hiring. It will be a much faster learning curve if they already have a list of tasks they need to perform. And it is well known that SOCs have a high employee turnover. This will also reduce pressure on senior SOC analysts as it will reduce the amount of time they need to spend on new SOC analyst education.

> **Important note**
>
> New SOC analysts should always undergo initial training and education as they need to learn about SOC processes and methodology. These technologies are here to help and not to replace important steps in SOC analyst onboarding.

Should SOC analysts perform only tasks written, then?

No, no, and no! Tasks and documented steps should only be used as guidance for SOC analysts on approaching incident investigation. However, attackers can change the way they perform attacks. If we don't perform deep analysis beyond the steps provided, it can give us false data that the incident is a false positive. At the same time, it may be a true positive as the attacker changed the method of attack that was not detected by documented steps.

Steps must be monitored and updated regularly. A common way to do this is either if tasks have the option to leave comments or via lessons learned. Either method is valid, but only if we have someone in the organization who will monitor those and apply changes as soon as possible.

If we will be making static tasks that are changed from time to time and not regularly, it is important to ensure that SOC analysts understand this and that they don't lean on tasks as the main investigation steps. This is something that can potentially be problematic for the SOC as we will have tasks that are not valid and can lead to confusion.

So, if the question is whether to use tasks or not – I would always be for it, but only if they are seen as guidance and if they are maintained regularly. If not, sometimes, it is better not to have tasks so that they don't lead to confusion inside the SOC.

Up until now, we have looked at the major aspects from an incident management perspective to help us manage incidents. However, an important aspect of the SOC is to investigate those incidents as well. That is what we will focus on next.

Investigation starting point – incident investigation page

At the beginning of this chapter, we discussed incident management and the incident queue, which contains a list of security incidents that have been detected. Those incidents contain specific information, such as ownership, status, severity, events leading to incident detection, and so on. We also mentioned that it is essential that the incident queue has a clean UI that is easy to read and navigate and doesn't contain much data. The SOC analysts must only be able to see the most important details of the incident on the incident queue.

But what about when a SOC analyst needs to see more data? How can SOC analysts investigate the incident?

For this reason, we must have an incident investigation page with more detailed information about the incident. In this view, we should be able to drill into the incident and investigate it.

OK, but isn't it easier to have it all on one page? The main UI goal is to reach the final point in the fewest clicks.

That is a valid point, but as we discussed beforehand, if we put too much information into one view, it will become too crowded, and it will be hard for SOC analysts to detect important information. Primarily, this is the case if a SOC analyst is performing an investigation using a smaller screen. We can see more data about specific incidents with a dedicated incident investigation page.

But wait, isn't this still incident management? Aren't we now speaking about the investigation?

Yes! The whole purpose of SOAR is to perform better and faster incident management. While the incident investigation page is part of incident management, this is a specific page from which we can start investigating more details. On the incident investigation page, we will have a list of events that are detected and that lead to incident creation. We will also have a list of essential entities such as IPs, URLs, hosts, and accounts. We will also have more information about incidents based on AI and ML, which are important aspects of incident investigation.

The incident investigation process

Having an incident management solution with space for investigation in it is a big plus for SOCs, and expanding it with other SOAR elements such as automation and reporting will certainly make SOC easier. But those tools must be used properly. In the investigation space, it is important to have a methodology on how to perform incident prioritization and investigation.

So, how do we start?

Execute incident prioritization

First, we need to be able to choose the incident we want to investigate. In the same time frame, SOC analysts will typically have a dozen incidents assigned, and it is vital to know how to *choose your battles*. This can be done in multiple ways, all of which are included in the incident queue:

- Start with severity. A higher severity will usually indicate a potentially more serious issue. If you have five incidents with low severity and one with high severity, the high-severity one should be the starting point.

- The next consideration would be to check what kind of entities are involved. If we detect incidents occurring in our domain controller or domain admin, this should take higher priority than a standard machine in the sales department.

- The incident type can show us a lot of valuable data. If it is privilege escalation or remote code execution, this can indicate a possibly more severe breach.

- Taking the whole picture into account can also help, which means seeing what kinds of incidents were detected before/after the incident we are targeting for investigation. Maybe there is a cybersecurity kill chain that we can detect.

- Nowadays, it is a standard that SIEM/SOAR solutions are connecting MITRE **Adversarial Tactics, Techniques, and Common Knowledge** (**ATT&CK**) tactics and techniques to incident detections. If this is the case, this can be another helpful step when prioritizing incidents because MITRE is one of the most well-known knowledge bases of adversary tactics and techniques used in the real world. With MITRE, we can build a cyber kill chain (how the attacker got access to the system, how the attacker moved through the organization, how privilege escalation was performed, and so on). This can also help us build an attack story and understand the common tools that are used for these techniques, as well as the common next steps so that we can perform a better investigation.

Once we have chosen our incident, we should start by understanding the events that lead to incident detection.

Conduct incident triage

Events that occurred before incident detection contain important raw data that can shed light on what happened and why the incident was triggered. It is essential to be aware of entities that are detected and to be able to quickly determine whether there is any HVA or VIP entity in it. This is important because if there is an HVA and VIP, there is probably a bigger blast radius and impact that a breach can have. As we mentioned earlier in this book, it is not the same impact when the incident host is a reception laptop and a financial server is involved. Based on this, we would maybe even want to lower or raise the severity level of the incident.

We also might want to perform additional enrichment of the incident. This enrichment can be to get TVM data of the host and check whether the IP is malicious or not based on external services such as VirusTotal or RiskIQ, user risk raised, and so on. This information can provide us with additional background information to understand whether the incident is a true or false positive, as well as whether there is a breach, how it started, and where.

Now, we should have a bit more understanding of the incident and what it is. It can be that the incident is a false positive as the IP is known and not malicious, or the user performed a specific action. However, it can also be that we still don't have enough information to determine whether this incident is a true or false positive. In this case, we would want to search for similar incidents to see whether we can get more background information. Suppose we followed the NIST or SANS incident management framework, especially the last step around lessons learned. In that case, we could read lessons learned from similar incidents to get more information to determine the incident's nature. Some SOAR tools use AI/ML and perform a similar incident search for you, and you can access those similar incidents from the incident investigation page. This is useful as it will save SOC analysts valuable time, which can be spent on more critical tasks.

Dig deeper for better context

Triaging incidents is only the beginning. With the additional information we get by performing triage, we can get a better picture, but not the whole picture of the incident and its correlated events. In most cases, we have to dig deeper to understand the true behavior of the incident.

Many SOAR tools have one more ace up their sleeve: the investigation graph. The investigation graph provides SOC analysts with a UI interface to connect the dots between different events. This is commonly connected to entities such as IPs, URLs, hosts, and so on. An investigation graph will allow you to find correlated events that may not be part of the incident you are investigating, which can lead you to detect the attack kill chain. Essential features in the investigation graph are to add those events to the incident, as well as the entities we find when performing the investigation in the investigation graph. This is important as it will give SOC analysts a better incident timeline and valuable data if an incident needs to be escalated to higher tiers.

But this still doesn't mean we can say whether it is a true or false positive. Sometimes, we must hunt through the logs to find more information in raw data. Tiers 2 and 3 mainly perform this step in incident investigation. It takes experience and more profound knowledge to understand those raw messages and know where to search for them. An important feature here is adding those raw events to the incident and seeing them in the timeline. This will help us see the whole picture of events connected to the same incident.

SOC analysts must also be able to respond to the incident. This can isolate the host, mark a user as compromised, add the IP to the firewall block list, and mark the host or user as *not risky* in our system. This is done via automation and by running different playbooks on the incident, event, or the specific entity in the incident. We will cover more about automation in *Chapter 3*.

Don't reinvent the wheel

There are a few features and processes that can help us in this process. Some of them are listed here:

- Have a defined incident management framework since having a structured process is the foundation of the SOC

- Utilize incident tasks to help SOC analysts with guidance on how to perform incident investigation

- Utilize automation for incident enrichment and response since we don't want SOC analysts to perform the same actions repeatedly

- Use SOAR AI/ML features such as *similar incidents* to provide more context

- Always search for features that will improve the SOC analyst's response time

There is no need to reinvent the wheel – utilize all SOAR features that are there to help you in the investigation process. However, sometimes, we need to dig deeper into the raw logs to understand what is happening with the incident. That is where threat hunting comes into play.

Understanding threat hunting

Because of many factors present in the cybersecurity space, as we mentioned in *Chapter 1* (not enough cybersecurity experts, expensive and hard to hire, and so on), many organizations are either not performing threat hunting or are doing it sporadically. SOCs are focusing more on Tiers 1 and 2 and detection tools for investigation.

But isn't that why we are paying for security tools?

Yes, security tools can be improved significantly by adding AI and ML capabilities, merging security products that cover different spaces into one (as we mentioned, XDR), and many more great examples. But bad actors are always searching for ways to penetrate, so security tools see it as an expected behavior or action.

This is where threat hunters come into play. As we mentioned, they are experienced security professionals, and their main job is to keep up with the latest security trends and keep up with bad actors. With this, they can hunt for specific data in the logs to find anything suspicious. Once they detect something suspicious, threat hunters dig deeper to see whether it is a one-time anomaly or whether there are more related events. Suppose they can see that there is the possibility that these events are incidents that need to be investigated further; they will create an incident with all the necessary information and work with Tiers 1 and 2 to investigate and respond to a possible breach.

Therefore, the investigation must include easy access for hunting and ensure that it is interconnected with incident creation or incident update. Once an event has been found, the SOC analyst must be able to either create new incidents, as we mentioned previously, or update an existing one with new important information. This can help SOC analysts to respond to an incident much faster and more accurately.

We mentioned that it is tough to find cybersecurity professionals that can perform threat hunting, which is why we see a significant rise in **Managed Security Service Providers** (**MSSPs**) and threat expert offerings. Instead of that organization performing hunting, they can outsource this job to more experienced professionals. This is important, especially for smaller organizations that don't have enough resources to add this tier level. More prominent organizations, in many cases, use MSSPs and threat experts in addition to their internal threat-hunting teams so that possible breaches can be detected as soon as possible. Internal teams will have more substantial knowledge about the internal systems used and the known behavior of those systems. In contrast, external threat experts will have better insights into the latest IoCs and TTPs.

Summary

This chapter covered the first two elements of SOAR – incident management and investigation. Here, we covered important aspects of incident management and what features it should include, such as incident assignment, status, severity, the option to add a comment, and more. We also introduced the two most known incident management frameworks – NIST and SANS – which are guiding us to organize how we should approach incident management, investigation, and response.

As part of this chapter, we discussed how SOC analysts should start with the incident investigation. We started by introducing incident prioritization, and then we continued with the importance of incident enrichment, lessons learned from similar incidents, and using investigation graphs. In the last section, we focused more on threat hunting as one of those features that SOCs are not focusing on as much as they should.

In the next chapter, we will focus on the last two elements of SOAR – automation and reporting. First, we will focus on the main automation features that can improve MTTA and MTTR. After automation, we will go through reporting, TI, and TVM as the final elements of SOAR.

A Deep Dive into Automation and Reporting

The last chapter covered two SOAR elements – incident management and investigation. This chapter will continue to drill down into SOAR elements and focus on **automation** and **reporting**. With more and more incidents to investigate, SOC analysts are often under pressure to ensure that the MTTA and MTTR meet the organization's policies. If we also consider that many incidents are similar and that a SOC analyst needs to perform the same actions repeatedly, it reveals why automation is such an important aspect of SOC and why it is a SOC analyst's best friend.

After looking at automation, we will jump into reporting, including how it can help organizations perform analysis, and how we can utilize it to hunt through data. We will then wrap up this chapter by focusing on **Threat Intelligence (TI)** and **Threat and Vulnerability Management (TVM)** and how they can enrich a SOC's investigation with invaluable data.

In a nutshell, the following topics will be covered in this chapter:

- An in-depth view of automation
- An in-depth view of reporting
- TI and TVM – how important are they?

Let's begin by discussing automation and what it is in the context of SOAR.

An in-depth view of automation

One thing that we've mentioned a few times already, and that will be mentioned a few more times, is that one of the most critical SOAR tasks is minimizing the MTTA and MTTR. There is no better way to do so than by utilizing automation.

Automation is commonly implemented using playbooks. A playbook contains a list of actions that will be performed once it runs. An action can be, for example, getting more details about an incident, getting more information about specific data from external services, or sending a notification to a service.

Let's look at the example of an incident investigation with no automation. Once an incident is detected, an analyst has to perform an initial triage to see whether the incident is a true or false positive. Commonly, that will be performed by looking at the entities (IP, account, host, URL, and so on) and activities associated with the incident. For example, say a user is signing in from an unfamiliar IP. The SOC analyst must check with whom the IP is associated, whether it is expected for the person to be logging in from that region, whether the user is on vacation, and so on. If they find that the IP is potentially malicious and want their firewalls, secure web gateways, or EDRs to block it, typically, the SOC analyst will have to issue a ticket to the respective teams to perform those tasks.

Let's now take the same incident with the addition of automation. Once an incident is created, the SOC analyst can assign a playbook to get the IP location and info about the user's sign-in location. So, when the SOC analyst opens up the incident, that info will be awaiting them immediately. If they see that the IP address does not match the user's expected location, they can run another playbook to check whether the user is on vacation and send info to the user to check whether they performed the sign-in; it may be that the user is on a business trip. If the SOC analyst gets info that the user didn't perform the sign-in, they can run a playbook that will automatically block that IP in their firewalls, secure web gateways, EDRs, and so on.

This can save valuable time for SOCs in detecting threats on time and reacting to, containing, and eradicating threats as soon as possible. With fast-spreading threats such as ransomware, a fast reaction to the incident is crucial.

What should be automated?

So, what steps should we automate? Everything? Or only the initial triage?

This can involve making both straightforward and complex decisions. What is the easy decision when it comes to implementing automation in the investigation step?

The first use case that we must consider for automation is the everyday tasks that our SOC analysts perform once an incident is detected. In most cases, this includes providing more information about the incident, such as IP address geolocation, whether the URL is malicious or not, user risk status and blast radius, or host info such as associated **Common Vulnerabilities and Exposures** (**CVE**) or security recommendations from TVM. If we check the IP geolocation, an incident with an IP is created; that is a perfect example of a step that should be done automatically. We don't want our SOC analysts to perform this step for every incident created, as it will overwhelm them. It is much easier to perform it automatically once an incident is created and add the verdict about the IP, URL, and so on to the incident. That can be in the form of a tag, comment, or any other method that allows the SOC analyst to easily interpret the information.

The second use case can be incident synchronization with external ticketing systems. Many organizations have one ticketing system that they use for IT support tickets, and they also want security tickets to easily monitor ticket assignments. Instead of the incident owner creating this manually, it can be done automatically. Some SOARs even offer bi-directional synchronization between SOAR and **IT Service Management (ITSM)** solutions.

The next use case should be steps that can be done automatically in response to a threat. As we have used IP as an example throughout, we will use it here as well. Say we get info that an IP address is known to be malicious; we probably want to block any traffic to and from that IP address. This is an excellent example of automation, as we can run a playbook on the incident that will grab that IP and send it to the block list in our systems. It will not require opening tickets with other departments to block it but will be performed automatically.

But what are the difficult decisions when it comes to automating a step? We want to automate as much as possible, but when is it enough? What shouldn't we automate? This is as important a consideration as *what* to automate.

When creating an automation policy in our organization (and there should be a policy), we have to define what steps in the incident investigation must be done manually, that is, what enrichments or responses we don't want to perform using automation.

As we stopped short of incident response when discussing easy decision-making, we can use it as the first example of a hard decision. Some users and hosts are business-critical in organizations and should only be blocked or isolated following specific instructions by specific personnel. Organizations often exclude these users/hosts from any automation processes so they won't be blocked or isolated if the wrong automation is run or if SOC analysts are unaware of the user/host's importance.

Another example of an implementation that would be hard is the entire incident investigation cycle using automation – from enrichment to response. For some incidents, this can be useful – if the IP address is malicious, block it in the firewall and close the incident as a true positive – but this should only be done if we have a possible brute-force attack, not if we have a successful sign-in from a malicious IP. This must be investigated further before closing the incident. For some, this can be problematic, as upon getting information that the IP is potentially malicious, we might automatically block it. Still, that information can be a false positive, and the IP might turn out to be safe. By blocking it, we may be cutting off employees' access to organization resources, leading to more problems if that employee is considered a VIP user.

The last example I will present is hunting, which shouldn't be done using automation. Hunting requires specific skills and knowledge and is mainly performed by senior SOC analysts that know the system inside out. Automation looks at presets of data, while hunting is done to find a needle in a haystack. Although there are some valid applications of automation when performing hunting such as data summarization or trend analysis, they should be used only to help focus on where to hunt or what for, but not to perform hunting itself.

As we can see from previous examples, decisions about what to automate are critical. There must be a policy in place about what can be automated, what cannot be automated, and where we need to include approval requests. In *Chapters 6-8*, we will go through examples of setting up automation and how to utilize it in real-life scenarios.

Automation is used not only in SOAR but also in many other technologies and industries. One concern that is often raised is the connection between automation and the user – in this case, the connection between automation and the SOC analyst.

Automation versus the SOC analyst

I am often asked, "Is automation a replacement for the SOC analyst?" No – never! Automation works best with SOC analysts and is set up to work alongside them. It is utilized to perform everyday tasks previously done by SOC analysts so that they can focus on more critical tasks, including educating others and expanding the field. It is used to reduce the pressure on SOC analysts and give them time to *breathe*. As stated in *Chapter 1*, the SOC analyst job role has significant turnover, as many SOC analysts burn out after a few years. Security is one of the most demanding professions, so there are not enough SOC analysts on the market. That is why we have to lower the pressure, and automation is there to *save the day*.

But what about the future? Will **Artificial Intelligence (AI)** and **Machine Learning (ML)** be able to perform everything automatically?

AI and ML play a more prominent and significant role in everyday life, not only in security. But just as we utilize AI and ML to detect threats, bad actors utilize that technology to create better threats. Because of this, we must rely more on security experts to create detections and automation and search for anomalies in the system to detect zero-day threats.

But one thing is certain – modern organizations will not be able to fight bad actors without SOAR and automation. The amount of security data and number of incidents is rising daily, and it is impossible for even the best SOCs to analyze all incidents manually. That is why we have to invest time in building automation to help our SOC analysts perform their day-to-day activities with ease.

Utilizing the SOC analyst or user input for automation

The question is whether we should trust user input when it comes to automation. Or, why should we ask for user input?

Let's go back to our incident example, where a user signs in from an unfamiliar location. SOC analysts cannot always know which locations a user will sign in from. Maybe if there are 10 employees, it is possible to keep track of their location, but unfamiliar sign-ins are daily events in organizations with thousands of employees.

In this case, we would like to get more info from the user about the sign-in itself. We can assign a playbook that will send the user an email with a notification about the sign-in event and ask whether or not the user signed in themselves. We can use that input from the user and add it as a note to the incident so that we have additional information when triaging the incident. We can now use location enrichment from solutions such as VirusTotal to check whether a sign-in is malicious or not, compare it with TI data that we are collecting, and then add user insight so that that decision can be made with higher assurance.

Many security experts see this as bad practice because if a user is hacked and a bad actor controls an account, the bad actor will likely have control over their email and be able to confirm that the sign-in is safe. If the SOC analyst only relies on user input, we can see how that could be problematic. This is why user input shouldn't ever be used as the only piece of information but only as additional information to correlate with other sources.

But getting feedback is essential as, for example, users can be on business trips, and the SOC analyst might not have information about it. User input is a great way to get this info. Many SOARs will offer you the option for the user to confirm the sign-in as safe and add a reason (business trip, working remotely, and so on); you can utilize the playbook to contact the user's manager to confirm the user's justification so that you have that additional level of assurance. But still – this info should only be correlated with other sources, not used as the key information in making a decision.

But what about SOC analyst input in the middle of a playbook run?

This is a handy feature for SOCs, as inputs allow SOC analysts to interact with the playbook and use the SOC analyst's input to determine what to perform next in the playbook itself. We can offer a few choices to SOC analysts and, based on the SOC analyst's selection, perform a set of actions connected to that specific selection. This is done mainly with conditions (such as an `if-then` statement). We will go through examples utilizing this action in *Chapter 7*.

But we have to be careful when utilizing this option as well, as the playbook will pause the run until the selection by the SOC analyst is made. Some SOARs offer a time limit to wait for input; if one is not selected, the default selection is used. This can be good practice as we don't want to wait for SOC analysts' input for an extended period.

To sum up, utilizing input from both users and SOC analysts can be great practice. Still, we have to be careful when utilizing these options, as they shouldn't be the primary source of information when making a decision. Humans are prone to errors; wrong selections can be made, bad actors can have control over accounts, and so on. These are all things we must consider when deciding whether we should utilize user input in automation or not.

Pros and cons of automation

I stand firm by the statement that there are no cons to using automation. It's all about the pros. Cons only arise from using automation incorrectly. Like all other security systems, proper configuration and administration are necessary to ensure automation works to our benefit, not our doom.

We must ensure that the automation configuration follows best practices and gets the correct data. If we make an error in setting up automation, bad decisions can result from the automation run, impacting our incident investigation.

We have to examine automation runs from time to time, ensure that all runs are successful, examine all automation steps, and ensure that all steps receive the correct information. A common mistake made is setting up automation and leaving it as is. Nothing in the world is perfect, and from time to time, everything needs to be adjusted. Bad actors keep up to date with security solutions and their actions; they will test automation and try to find out how to *slip through the cracks*. Therefore, we have to monitor our automation scenarios so that they can continue to work to our benefit.

A key thing to note is that automation is used to help SOC analysts and that it can only work to their benefit if used correctly and, as mentioned, by following best practices. If you do that, you will get all the pros of automation and evade the cons.

Automation helps us to perform some important incident detection tasks, with no or limited input from SOC analysts, but we also need to see the impact of SOAR, to get a more statistical and visual representation. This is where reporting comes in.

An in-depth view of reporting

With a much clearer view of automation, it's now time to shift our focus to reporting and its role in SOAR. Reporting is not a new tool and is not strictly connected to SOAR. All SIEM solutions have their own variation of reporting. In most cases, SIEM reporting is more about visualizing the huge amount of data that is being ingested, while in SOAR, it's mostly about reporting on incidents, automation, and overall SOC performance.

While most SIEM and SOAR solutions are still separated into their own spaces, some SIEM and SOAR tools are integrated, such as Microsoft Sentinel, and SIEM and SOAR use the same reporting mechanisms.

We can expect this to be the case more and more, and the line between SIEM and SOAR will likely, and should, disappear. This unification of SIEM and SOAR will bring easier management and integration of services, such as reporting, where we will have better overview and management, as well as fewer portals for SOC analysts to visit.

In this section, I will focus more on the benefits of unified reporting (SIEM and SOAR) than SOAR itself. Why? Because it's much better. Instead of separating reports and then joining them later on, you can compare raw data in SIEM with performance data in SOAR.

This will also give us an edge to investigate incidents from statistical data, not only from incidents themselves, and we can detect patterns in statistics that we might miss in the sea of incidents. But then, in those reports, we can also utilize automation capabilities to perform some additional enrichment, or even to provide a response, to the statistical data we found.

Let's first learn what reporting is.

What is reporting and why is it crucial?

If we have implemented the other elements of SOAR, such as incident management, investigation, and automation, why is reporting also required? Is it just another way for managers to micro-manage?

To be honest, we can use it just like we'd use reporting in any other context, but that is not the main idea behind it. The idea is that reporting in the context of SOAR is hassle-free for SOC analysts, and reports are automatically filled in with data available in SIEM/SOAR tools.

Reports are meant to be a visual representation of data, making it easy to digest, understand, and navigate through. Usually, SIEM's data is stored in tables and contains hundreds, thousands, or, in many cases, millions of records per day. We cannot see statistics and common occurrences in the logs just by looking at this raw data. It's much harder to follow changes over time and how data and actions impact SOCs.

For example, in SOCs, one critical report is about the number of open incidents, how many of them are not assigned yet, how many have owners, what the prior status and comments of each are, and so on. Another excellent example for SOCs is measuring the MTTA and MTTR over time and how specific policies they implement will affect those numbers.

These two examples are great showcases of how reporting is meant to help SOCs, how those reports are created from existing data in SIEM and SOAR solutions, and how reporting doesn't need input from SOC analysts. SOC analysts need to ensure good standards of SOC hygiene are maintained – closing incidents with the correct classifications, moving incident statuses, ensuring comments are kept up to date, and ensuring that bulk jobs are not done once a day (or even once a week) – but regular updates are carried out with actual information as it is happening.

If all this is filled out correctly, the SOC manager can have all the info at hand without asking SOC analysts for additional information in 99% of the cases. That is important, as having to fill out individual report cards would take the time of SOC analysts away from focusing on incidents.

But reports can also be used for investigation purposes. Utilizing parameters in reports can help organizations create dynamic reports based on what they choose to focus on in each reporting section. So, for example, say we want to see a list of users who failed to sign in. Then, from that list, we select one of the users to see from what locations they have failed to sign in. Then, we might see a country that we don't expect users to sign in from, and by selecting it, we can get details about that sign-in. This is done without writing queries but by drilling down through reports. Therefore, reports are crucial for SOC managers and SOC analysts to perform swift investigations or get more details about specific scenarios.

There are many positive elements of utilizing reports in day-to-day SOC life as they can be a time-saving tool, especially when we need to visualize a more significant amount of data, search any statistics around it, and see how data is behaving across a specified period of time. If we see that the number of failed sign-ins for a user is 3 per day at first, and it increases to 10-15 per day, that can be a sign that someone is trying to breach an account and stay undetected from the system. This is one sign that we should investigate this user further.

Are reports the new incident management?

Reports are a powerful tool, and we can create many things with them. I have witnessed companies creating their own incident management queues using reporting, but should we do this? Is that the way to go? That question requires the same answer as "What came first – the chicken or the egg?"

Incident management solutions are robust and give more and more functionalities as they grow, but there is no such thing as one size fits *all* – only one size fits *most*. SOAR vendors must focus on the features and possibilities that most companies are asking for and will benefit from when developing features, but some organizations will have specific use cases that will not be part of SOAR, or they will migrate from one tool to another and the previous tool may have had a feature that they previously utilized that the new solution doesn't have.

One thing that organizations can do is work with SOAR vendors on implementing that feature, but this can take time. Vendors work according to prioritization, making enough resources to develop that feature, testing it out through the system to ensure it works, and so on. Should we create and utilize it through the report itself? If we have a person proficient at creating reports, why not? This report can be a fantastic resource you can utilize until that specific feature is developed, tested, and put into production by the vendor. The great thing about SOARs is that it has communities and many templates that you can utilize. If something similar has already been created, we can utilize it and fine-tune it for our needs.

But this should only be used for really specific requirements and situations. Incident management solutions and reports are divided for a reason, and you can miss a lot of native incident management features by creating a new incident management solution from reports, as reports will not have the same look, feel, and possibilities as native incident management solutions.

Typically, SOCs will not want to shift SOC analysts from incident management to reporting, back to incident management, and so on. We want our SOC analysts to make as few clicks as possible and utilize as many native features as possible, which will give users the best experience in the long term. In the short term, reports can help, but in the long run, they can have more downsides than benefits.

What is the proper way to utilize reporting?

As we mentioned, reporting can end up being used as a micro-managing tool. This usually ends up being the case because reporting is not utilized in the proper manner. Hence, here are a few use cases when reporting should be used:

- **Incident report**: This is a report integrated into SOAR that will give a fast incident summary with all the important information around the incident (a description, the severity, tactics detected, automation runs, comments, entities/artifacts, incident audit data, and so on). If the incident report isn't integrated, then it can be created from scratch. The idea of this report is not to ask SOC analysts to send incident details over email for reporting purposes but for it to be generated in seconds with no SOC analyst interaction.

- **SOC performance**: This was already mentioned as a use case earlier, and it is really important to use the metrics – especially the MTTA and MTTR.

- **SLA breaches**: This is an important report, not for micro-managing SOC personnel but to follow up on statistics of the SOC. Some **Service Level Agreements (SLAs)** are easy to meet, but others are not as easy. Monitoring what incidents or tasks have the most SLA breaches can help us investigate why it is so and create better practices. This can be hard to detect per incident or task base, but if we look at the bigger picture and a longer period of time, we can start getting better results.

- **SOC statistics**: I recommend that this is a different report from the one on SOC performance. In this one, we should focus on the statistics themselves, such as how many of the same incidents we have, which incidents take the longest to investigate, what automation we aren't utilizing, and what tasks are not completed. From this, we can gather more information and use it to improve existing tasks and automation, remove unnecessary tasks or automation, create better practices for frequently occurring incidents, fine-tune them, and so on.

These are just a few examples, and many SOCs will have plenty more, but the idea is that for each of them, we utilize the data already in the system and limit SOC analyst involvement. SOC analysts should be there just to provide feedback on improving the process based on SOC statistics.

The preceding reports are not for SOC analysts but more for management. What reports might SOC analysts need? The main report that SOC analysts should have is one to drill down through all logs that are ingested into the SIEM tool. Hunting is an amazingly useful activity, but it requires a lot of knowledge. Reports can give even entry-level SOC analysts the power to look through a huge amount of data and find connections. For example, we might investigate an incident relating to a certain user by opening a report that summarizes the user's sign-ins over the last few weeks. Also, audit information about access, any previous incidents, and common incidents for that user, plus utilizing **User and Entity Behaviour Analytics (UEBA)** information to see peer information to compare to that user's, and so on, can give us a much better picture while conducting the investigation.

However, there is one report that I would like to urge SOCs not to use – analyst performance monitoring reports. This can be a great tool for training junior SOC analysts and even for advanced SOC analysts to use to show where improvements can be made, but this is one type of report that is never used for its intended purpose – more to micro-manage people. Not only this but the statistical data may also fluctuate for various reasons. SOC analysts may have many incidents assigned, or an incident assigned to one analyst might need more time for investigation than an incident assigned to another analyst. Also, there may be one-of-a-kind incidents where we don't have procedures, and so on. While all this data is hugely important to improving metrics, it is often misinterpreted as poor SOC analyst performance. So, if it were up to me, either the manager wouldn't be able to access this report, or it shouldn't be used. Statistics are great for showing the bigger picture, but they never show the whole picture. There are just too many variables.

With reporting, we have covered the last major element of SOAR. But there are two more common topics when speaking about SOAR – TI and TVM. In the next section, we will go over why TI and TVM are important, as well as some common use cases in which TI and TVM are relevant.

TI and TVM – how important are they?

TVM analyzes vulnerabilities within the organization and processes how to patch those vulnerabilities. In the world of SIEM/SOAR, we will speak mainly about CVE, which refers to vulnerabilities that can be found connected to our systems and machines and can be compared with additional information.

TI, or **threat intelligence**, is information that can point us to potential threats inside our systems. It can contain information about threat actors, tactics and techniques, observations about IP, hosts, URLs, and so on. TI information is collected, processed, and analyzed to help organizations understand what could be the possible next step attackers make, as well as what their final motive is if they have one. When we speak about SIEM/SOAR and TI, we are mainly referring to **Indicators of Compromise (IoCs)**, such as IP addresses, file hashes, and URLs. That information is ingested into the system and compared with other available data.

Typically, SOCs will want both pieces of information inside of their SIEM solutions so that they can compare TI and TVM data with ingested data. But why?

Let's recall when we spoke about using automation to get information about an IP address from VirusTotal. We wanted more information about that IP address to make faster and easier decisions when triaging incidents. VirusTotal is one of the TI services from which we can get information, and there are many more, but this can also be done by ingesting TI data directly into the system and utilizing that data inside the detection query itself. As most IoCs have a confidence rate, we can, for example, write in the detection query that if we have more than 60% confidence, that IoC is a true positive and an incident should be raised, or, if it is lower than 60%, not to raise an incident. With this, we remove the need for the SOC to investigate every incident, as an incident will not even be raised if it's unlikely to be a true positive.

TVM data can also be helpful when deciding which incident we want to investigate first. Let's imagine we have a few incidents that contain hosts, and we are making prioritizations. All incidents are of medium severity, and hosts are of the same value. Usually, we would try to see what kind of incident it is or go with the first-come-first-served method, but what if we have additional information about the machine that can help us? What if we can check whether there are any active CVEs in the machine and if we raise the severity of an incident or list CVEs in the incident comment/note space? That would help, right?

If we can see that specific machines have five active CVEs and two of them are critical, while other hosts have one or two low CVEs, we can make a decision on which one should be investigated first. Maybe the incident is connected to a public exploit connected to that certain CVE, and we maybe need to perform emergency patching in our system for that CVE.

With that, we have completed an in-depth overview of the SOAR elements, and for each element, I mentioned how crucial they are for SOCs, but is there one that can be considered to be at the top of the list, ahead of the others? I would say no – they are all essential for an effective and efficient SOC. You can focus on specific elements, but what makes SOAR so strong is the unity between these elements. Missing any of them can be problematic and lead to slower breach detection. It is similar to whether we would choose between eating or drinking water. If we only eat, we will die from dehydration. If we only drink water, we will die from hunger. The same is true with SOAR elements; you can choose one over the other, but that can lead to late breach detection, poor SOC performance, increase the load on SOC analysts, big SOC analyst turnover, and so on.

Summary

In this chapter, we went through automation and reporting and how they can help organizations minimize the MTTA and MTTR.

We discussed one of the most important aspects of automation: deciding what steps in our incident investigation should be automated. Then, we moved on to hot topics when thinking about automation. We first discussed whether automation is a replacement for a SOC analyst or their best friend. After this, we moved on to looking at using input from SOC analysts and users and how we must be careful when utilizing these automation features.

From a reporting perspective, we spoke about creating an incident management solution by utilizing reports and making it more customized to our needs, and that spending time on these tasks makes sense.

In the last section, we went through TI and TVM in SOAR and how they can play a significant role in an investigation. TI and TVM can influence decision-making in tier 1, which was covered in *Chapter 2*, and directly impact the MTTA and MTTR in organizations.

The next chapter will go through some of the most well-known SOAR tools, such as incident management, investigation, and automation, and discuss how these features are utilized in day-to-day SOC life. These solutions are now mostly part of more comprehensive SIEM tools, but we will explain how those SIEM tools were given an additional push as a ruling security solution.

Part 2: SOAR Tools and Automation Hands-On Examples

As the first part provided a more theoretical perspective, the second part will focus on real tools and examples. At the start of Part 2, we will introduce a few SOAR tools and showcase what SOAR elements look like in real life. After that, we will focus on automation using Microsoft Sentinel, where we will create playbooks step by step – from planning and designing to testing them.

This part contains the following chapters:

- *Chapter 4, Quick Dig into SOAR Tools*
- *Chapter 5, Introducing Microsoft Sentinel Automation*
- *Chapter 6, Enriching Incidents Using Automation*
- *Chapter 7, Managing Incidents with Automation*
- *Chapter 8, Responding to Incidents Using Automation*
- *Chapter 9, Mastering Microsoft Sentinel Automation: Tips and Tricks*

4

Quick Dig into SOAR Tools

The previous chapters introduced SOAR as a tool and discussed how it can help in day-to-day SOC operations. It can start with case management, helping to orchestrate incident assignments. Then, it can automate everyday tasks where SOC analysts use many tools/windows, enrich an incident with additional data, or even respond to the incident. Finally, it can assist with reporting for better analysis and incident response planning in the future.

In this chapter, we will focus on a few popular SOAR tools and understand how they are combined with SIEM tools. We also dive deep into their main functionalities and learn how they can be used for incident management, investigation, automation, reporting, TI/TVM, and administration panes.

The tools that will be covered are as follows:

- Microsoft Sentinel SOAR
- Splunk SOAR (Phantom)
- Google Chronicle SOAR (Siemplify)

Some other SOAR tools include Palo Alto Cortex XSOAR, IBM Security SOAR, Swimlane SOAR, LogicHub SOAR, and so on. Because the functionalities offered by SOAR tools available on the market are mainly similar, we will not cover all of them in this chapter.

> **Important note**
>
> This overview is achieved by utilizing publicly available documentation and limited access to the tools. To learn more about each product, please consult with experts on each of the products. This is only an overview of the solutions, with the features available at the time of writing (October 2022). The tools/features described in this and the following chapters might change or be enhanced in the future.

Microsoft Sentinel SOAR

Microsoft Sentinel is a cloud-native SIEM and SOAR solution. It utilizes the power of Microsoft Azure to scale SIEM/SOAR demands, even for the biggest customers. Microsoft Sentinel collects data from various systems such as first-party Microsoft solutions, Syslog, **Common Event Forwarding** (**CEF**), and **Application Programming Interfaces** (**APIs**). All the data collected is stored in a Microsoft Azure solution called **Log Analytics Workspace** (**LAW**). Microsoft Sentinel is enabled on top of LAW and is directly connected to it via a relationship – Microsoft Sentinel can be connected to only one LAW, and vice versa.

Microsoft Sentinel utilizes incidents and alerts for detection. Incidents are the primary investigation mechanism, while alerts can be seen as evidence that some incident has happened. One incident can contain between 1 and 150 alerts. The detection mechanisms in Microsoft Sentinel are as follows:

- **Scheduled analytic rules**: These are written using **Kusto Query Language** (**KQL**), and you can choose to write your own or utilize existing ones from rules templates or **Content hub**. Scheduled rules can be configured to run anywhere between 5 minutes and 14 days.

- **Near-real-time (NRT) analytic rules**: These are written in KQL as well, and the main difference is that NRT detections are run every minute; thus, they look at data ingested even at the last minute.

- **Fusion detection**: This is based on ML algorithms and is used to detect multistage attacks.

- **Microsoft security incident creation**: This involves creating incidents based on alerts from other Microsoft security solutions.

- **Anomalies**: This involves using ML to detect anomalous behavior. Anomalies are not used to trigger alerts or incidents but can be used for better correlation and investigation.

All incidents and alerts are written in their corresponding tables in LAW, called `SecurityIncident` and `SecurityAlert`.

Let's assume you want to send alerts from different tools to Microsoft Sentinel SOAR. If so, you will need to ingest them into a Microsoft Sentinel table and then create a detection rule in Microsoft Sentinel for that alert or incident. This process is slightly different from standard SOAR tools, as Microsoft Sentinel SIEM and SOAR are integrated, while most other vendors have SIEM and SOAR separated.

In the next few subsections, we will go over the SOAR elements in Microsoft Sentinel. We will introduce the essential features and describe how to utilize them.

Incident management

The first element that we will dig into is incident management and its main functionalities in Microsoft Sentinel. All incidents created in Microsoft Sentinel can be found under the **Incidents** tab. It will, by default, show incidents detected in the last 24 hours, but you can filter to any time between now and 90 days.

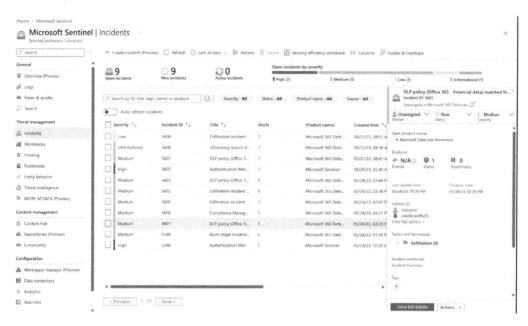

Figure 4.1 – The Microsoft Sentinel Incidents page

For each incident, we can see important information such as **Title**, **Severity**, **Product names**, and how many alerts are on the incident. You can filter which columns you want to check out by selecting **Columns** in the top menu.

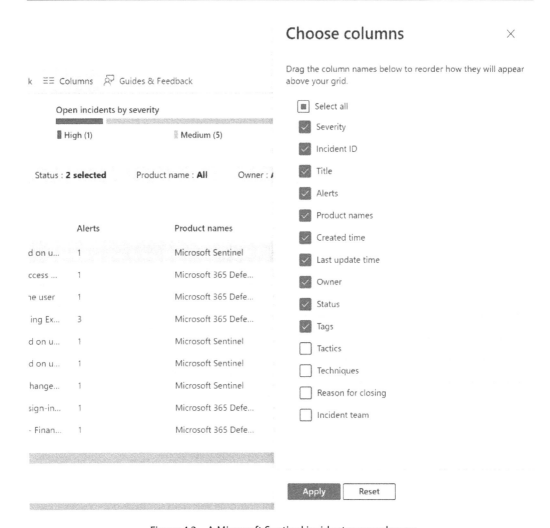

Figure 4.2 – A Microsoft Sentinel incident page columns

Clicking on each incident will show you basic information about the incident itself. The first page will allow you to search for specific incidents based on criteria. A basic search can include the incident title, tag, owner, ID, or product name, or you can use an advanced search based on the alert, entity (account, host, IP address, or URL), bookmark, and so on.

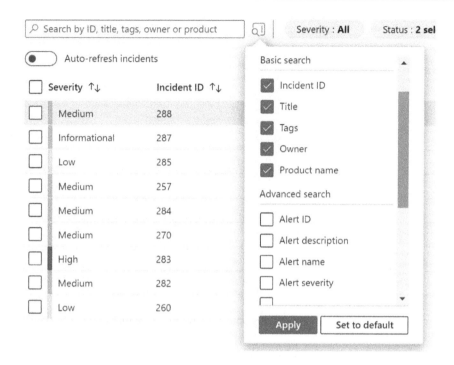

Figure 4.3 – Microsoft Sentinel incident page search

There is also the option to filter incidents based on *severity*, *owner*, *status*, and *product name*.

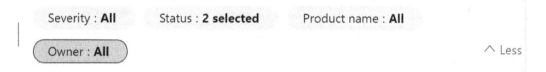

Figure 4.4 – Microsoft Sentinel incident page filtering

When we select one of the incidents, it will open a small window on the right side of your screen with primary data on the incident. This data will include the incident title, the owner, the status, the severity, a description, entities, the number of events, alerts, and bookmarks. From this view, it is possible to perform some immediate actions. For example, from the top menu, you can assign an owner and change the status and severity. From the bottom menu, you can open a full incident view or perform an action, such as opening an investigation graph, running a playbook, creating automation rules for future detections, or creating a team in Microsoft Teams for collaboration on the incident.

Figure 4.5 – Microsoft Sentinel incident page details

We can also select multiple incidents, allowing us to perform bulk options such as deleting incidents, changing severity or owner and status details, and adding tags.

Let's suppose we need to create an incident manually. If so, we can do it from the incident page or utilize API calls and/or any playbook already available, such as creating incidents by filling in a Microsoft Forms form or sending emails to shared mailboxes.

Investigation

After identifying incidents, we need to perform an incident investigation. In Microsoft Sentinel, we will do that using the incident investigation experience.

When we select an incident and click on **View full details**, it will open a new page where we can perform a more in-depth investigation.

Important note

Microsoft Sentinel had a major incident investigation experience GUI update in January 2023. This chapter is written per the GUI in October 2022. At the time of writing, you could change to a previous incident investigation experience if you need it for reference.

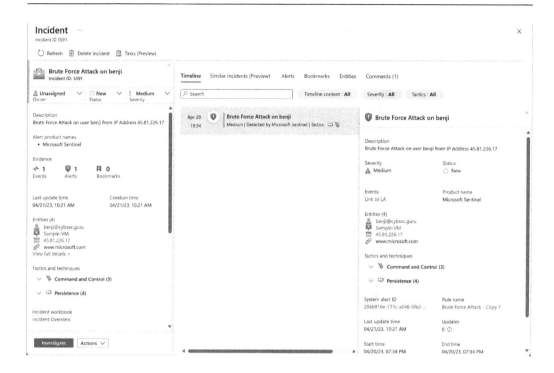

Figure 4.6 – The Microsoft Sentinel incident queue page

In this view, we have options like before, but we also can see a timeline of events, alerts, bookmarks, entities, comments, and similar incidents. **Similar incidents (Preview)** provides a handy list of incidents with similar entities or detection rules, including the one we are investigating. We can open each of these incidents to check for more information.

Figure 4.7 – The Microsoft Sentinel Similar incidents (Preview) tab

We can also run a playbook on an incident, run a playbook on an alert, create an automation rule for this specific incident, or create a team on Microsoft Teams for better communication on the incident itself.

When creating a Microsoft Teams team, we can immediately add people we want to work with, and once an investigation is done and an incident is closed, the team created on Microsoft Teams will be archived automatically.

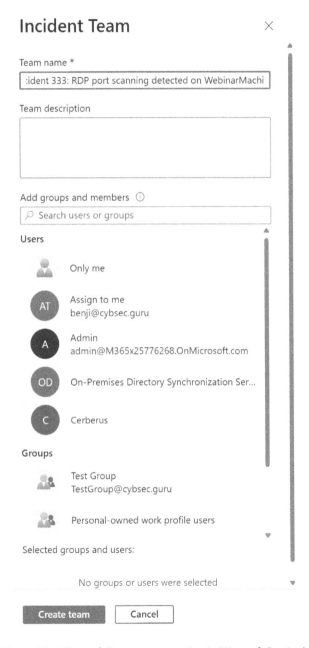

Figure 4.8 – Microsoft Teams team creation in Microsoft Sentinel

We can also utilize an investigation graph that allows us to find similarities between alerts in the incident we are investigating and other alerts, based on the entities found in the incident's alerts and other alerts detected on Microsoft Sentinel. If we find an alert that we think is part of the incident we are investigating, we can add it to the incident from the investigation graph. We can also see more data on the right-hand menu, such as a timetable of all alerts in the investigation graph, entities, and insights based on UEBA.

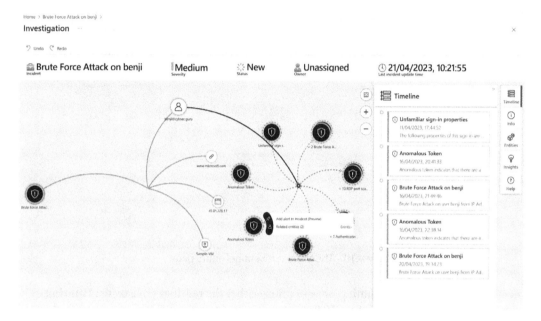

Figure 4.9 – A Microsoft Sentinel investigation graph

Let's assume we deduce from this investigation graph view that an IP address, URL, or file hash we have as an entity is malicious, and we need to track it. In that case, we can add that entity to the **Threat Intelligence** blade, which will also be written in the `ThreatIntelligenceIndicator` table in Microsoft Sentinel. Then, we can use it for hunting, creating new detections, and so on.

Each entity in Microsoft Sentinel has its own page that contains all vital information about it. It can be enriched by UEBA information such as entity insights, data based on Azure AD, and geolocation information.

Figure 4.10 – The Microsoft Sentinel entity page

If we need to perform threat hunting, we can utilize either the raw logs table or the **Hunting** or **Notebooks** part. **Hunting** is based on KQL queries and provides a space where you can save them, easily access them, and run them. From the hunting results, we can create bookmarks and assign them to the incident we are investigating.

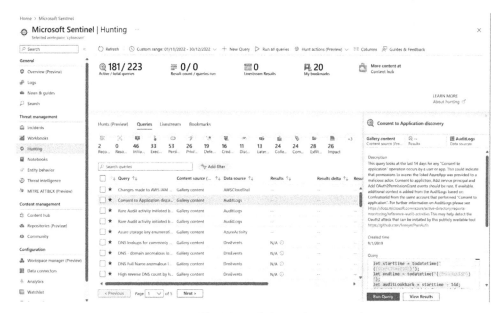

Figure 4.11 – The Microsoft Sentinel Hunting tab

Notebooks are based on Jupyter Notebook and utilize Azure ML to hunt through the Microsoft Sentinel logs. **Jupyter Notebook** is an open source tool used for the creation and sharing of documents that contain code, text, and visualization.

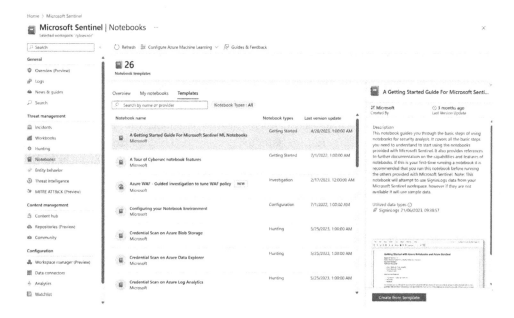

Figure 4.12 – The Microsoft Sentinel Notebooks tab

Next, it's time to go over automation and how we can automate steps that SOC analysts perform to improve MTTA and MTTR.

Automation

Microsoft Sentinel utilizes automation rules and playbooks for automation, which can be accessed from the **Automation rules** and **Active playbooks** tabs.

Figure 4.13 – The Microsoft Sentinel Automation tab

Automation rules are the central management of Microsoft Sentinel automation. They can be triggered on incident creation, update, or alert creation. After the trigger, we configure **Conditions** when the automation rule runs, as well as **Actions** that the automation rule will perform. Native actions include assigning owners, changing severity or status, adding a tag, and running a playbook. All automation rules have order numbers, which are run from first to last and one by one.

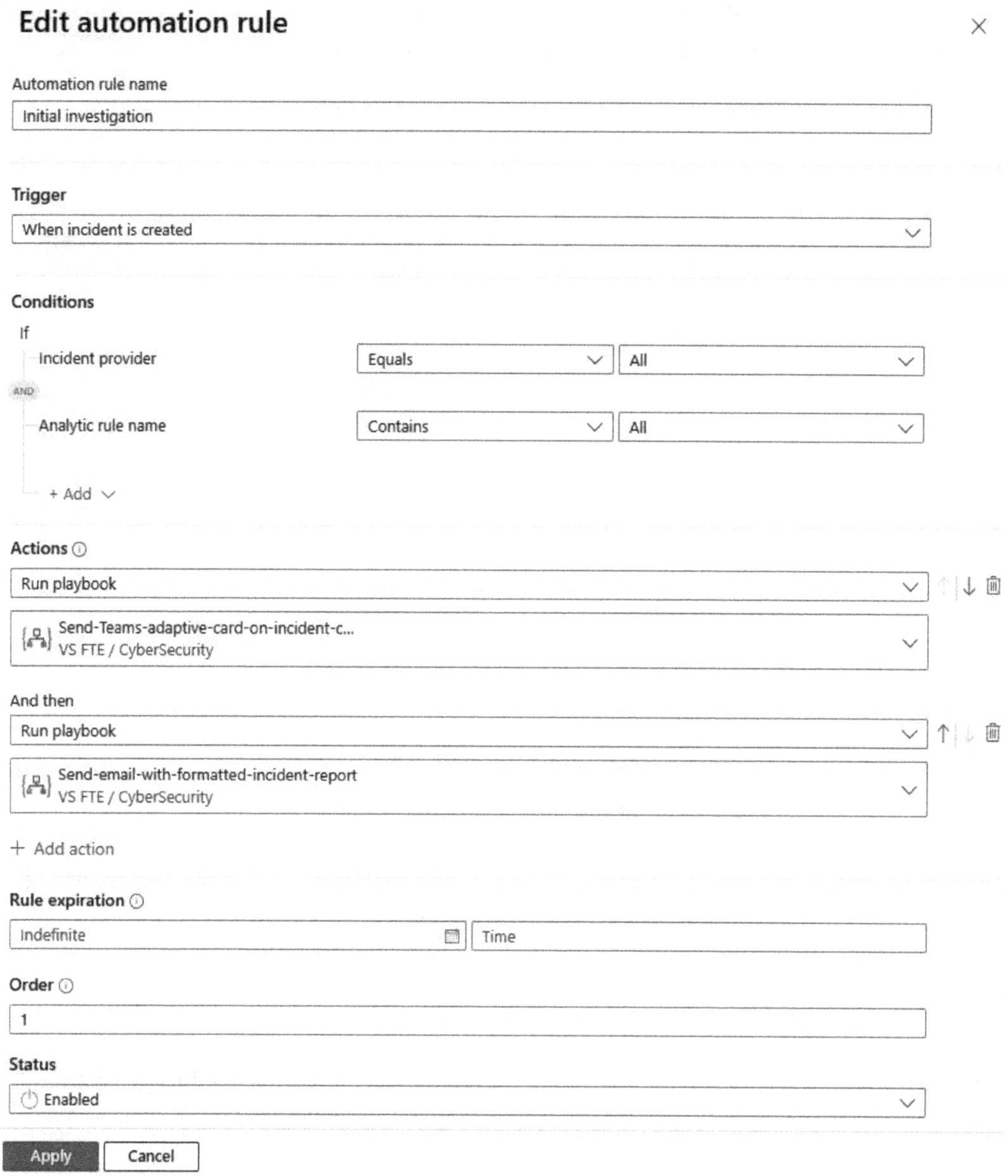

Edit automation rule   ✕

Automation rule name

| Initial investigation |

Trigger

| When incident is created   ⌄ |

Conditions

If

| Incident provider | | Equals | ⌄ | All | ⌄ |

AND

| Analytic rule name | | Contains | ⌄ | All | ⌄ |

+ Add ⌄

Actions ⓘ

| Run playbook   ⌄ | ↑ ↓ 🗑 |

| 🖳 Send-Teams-adaptive-card-on-incident-c...
VS FTE / CyberSecurity   ⌄ |

And then

| Run playbook   ⌄ | ↑ ↓ 🗑 |

| 🖳 Send-email-with-formatted-incident-report
VS FTE / CyberSecurity   ⌄ |

+ Add action

Rule expiration ⓘ

| Indefinite   📅 | Time |

Order ⓘ

| 1 |

Status

| ⏻ Enabled   ⌄ |

[Apply]   [Cancel]

Figure 4.14 – Microsoft Sentinel automation rule creation

Playbooks are a list of actions that will be performed automatically when they are triggered. Microsoft Sentinel has two triggers – *incident* and *alert* triggers. Microsoft Sentinel utilizes a Microsoft Azure solution called **Logic Apps** to create playbooks.

Figure 4.15 – Microsoft Sentinel playbooks creation

In the next chapter, we will dive deep into Microsoft Sentinel automation. But for now, we will go through reporting in Microsoft Sentinel connected to SOAR.

Reporting

Microsoft Sentinel uses **Workbooks** as a feature of Azure Monitor and LAW to create reports. All reports are written in KQL, which reads data from the SIEM part of Microsoft Sentinel. As Microsoft Sentinel SIEM and SOAR are integrated, we can use native SIEM capabilities with SOAR, and vice versa.

One of the workbooks integrated into the incident page is **Security Operations Efficiency**, where you can see all data about the SOC. As all workbooks have the option to be edited, you can edit workbooks per your needs or even create new workbooks for your day-to-day operations. Many workbook templates can be found under **Workbooks** and in the **Content hub** blade so that you can utilize pre-created content.

Figure 4.16 – Microsoft Sentinel workbooks

Now, let's walk through the TI and TVM elements of Microsoft Sentinel SOAR.

TI and TVM

Since Microsoft Sentinel has SIEM and SOAR integrated, it also provides SIEM TI benefits directly to SOAR. In Microsoft Sentinel, we have a **Threat Intelligence** blade where we can find all our TI indicators; it also provides us with the option to manage TIs. All TI indicators are written in the `ThreatIntelligenceIndicator` table in Microsoft Sentinel so that we can use them further for detection, hunting, or reporting.

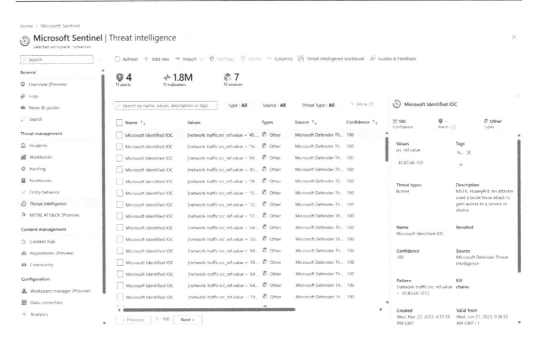

Figure 4.17 – The Microsoft Sentinel Threat intelligence tab

To get TI data into Microsoft Sentinel, you can utilize one of the following options:

- Integrate a **Threat Intelligence Platform** (**TIP**) product
- Connect to a **Trusted Automated eXchange of Indicator Information** (**TAXII**) server
- Utilize the Microsoft Sentinel TI indicators API

Regarding TVM data, you can ingest that data into Microsoft Sentinel and create detections, hunting queries, and reporting mechanisms around it. An example of TVM data that can be brought to Microsoft Sentinel is TVM advanced hunting data from Microsoft Defender for Endpoint.

Splunk SOAR (Phantom)

Splunk is a well-known name in the SIEM field and, in contrast to Microsoft Sentinel, can be installed in your local infrastructure (also called *on-prem*) and also utilized in the cloud, using **Amazon Web Services** (**AWS**), **Google Cloud Platform** (**GCP**), and **Microsoft Azure**. Splunk can ingest data from various sources, including local environments such as servers and firewalls, and collect data from cloud infrastructure.

Splunk SIEM can create alerts based on searches for historical and real-time data.

The Splunk SOAR component is an additional service that must be configured and connected to Splunk SIEM. Those who have worked with SOAR solutions for years will mostly call Splunk SOAR *Phantom*, as Splunk bought Phantom SOAR back in 2018 and integrated it as an internal SOAR solution.

Splunk SOAR, the same as its SIEM solution, comes in two versions that can be installed in local infrastructure or the cloud. Splunk uses Splunk Apps to connect different sources, and those applications are also used in SIEM and SOAR solutions. Splunk SOAR can connect to other SIEM solutions using its Splunk Apps, where you can collect alerts, incidents, and events.

Splunk SOAR utilizes *events* and *cases* for investigations. Events are detections made by SIEM or other sources from where detections are imported. After initial triage, you can add detection to new or existing cases. If we connect parallel to Microsoft Sentinel, detections are alerts, and cases are incidents.

To connect Splunk SOAR to any solution we want to use for incident management and automation, we need to utilize Splunk Apps to make connections. Many pre-created apps for connection can be found in Splunk Base.

Next, let's look at Splunk SOAR incident management and investigation and understand its main elements and functionalities.

Incident management and investigation

The first page in Splunk SOAR gives you an overview of SOAR space, and you will see open events, your total workload, top playbooks and actions, the health of the asset, whether there are any pending approvals, SLA breaches, as well as an automated **Return of Investment** (**ROI**) summary. These widgets are configurable, and you can remove or collapse any of them. Also, you can filter them by sources or users.

Figure 4.18 – The Splunk SOAR Home tab

The **Events** tab can be found under **Sources**, and you can filter events by assigning them to you as an analyst or filter by new events or those with **High** severity. You will also get the option to search for events based on the event name or ID.

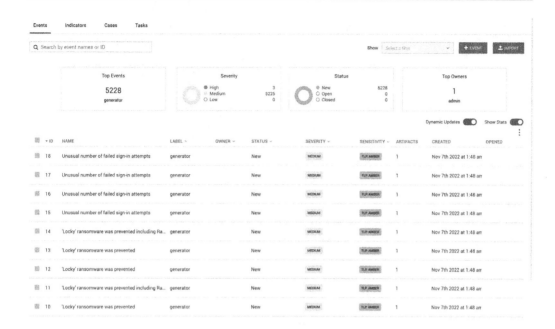

Figure 4.19 – The Splunk SOAR Events tab

Each event can be opened for initial triage, and you will have investigation space for each. The first view will give you a timeline of events, where you can see when an activity started and how artifacts and evidence are added, and you can select each to see more details. For each event, you can run actions and playbooks, edit and change status, severity, and SLA, among others, as well as promote an event to a case.

Figure 4.20 – The Splunk SOAR event view

Events have support for workbooks, which, in Splunk SOAR, are packaged with tasks on how to investigate and mitigate events and cases. A workbook can be added manually to events but also when promoting an event to a case. Those tasks can contain playbooks and actions that you can run directly from a task. Tasks assigned to a user can be seen from the **Tasks** view under **Sources**.

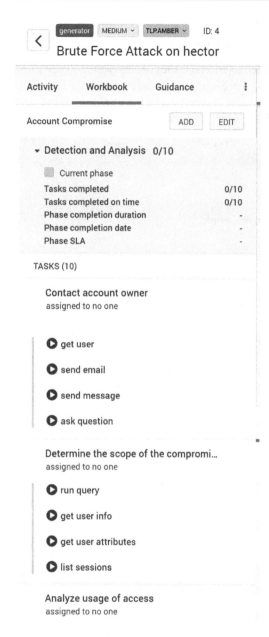

Figure 4.21 – The Splunk SOAR workbook

In the **Activity** tab, you will see information about activities completed as playbooks run and whether there is any approval waiting as part of a playbook. This is also a space where we can add comments, mention other users in comments, utilize commands to run a playbook or action, and so on.

The **Guidance** tab will provide information, such as mission experts, that you can add to collaborate on an event, playbooks to run, or actions to run.

If any files need to be uploaded to an event, they are supported inside **Files**, but you can also create an event report if it needs to be presented to management. The event report will include all details and actions from the event.

Figure 4.22 – The Splunk SOAR event report

Once an incident is promoted to a case, you will be able to see it under **Cases**. Cases can have multiple events and can be added as artifacts or evidence. The rest of the case management is the same as with event management. It's important to note that with an artifact, you can edit it, remove it, add it to another case, and run a playbook on it, but evidence can only be seen and removed.

Automation

Splunk SOAR utilizes *actions* and *playbooks* for automation.

Playbooks can be selected in two ways – to be run as automation, where a playbook can be used as a standalone or sub-playbook, or as an input that will require additional input to run it. When we utilize an input playbook, it can be used only as a sub-playbook.

Once you select a starting point in the playbook configuration, you can add the following to the playbook:

- **Action**: This is a specific action we want to perform based on the connector we add for a specific source. For example, if we add an Azure AD connector, we can block a user or reset their password.
- **Playbook**: A possible next step after running a playbook can be to run a different playbook from our list (a sub-playbook).
- **Code**: We can run code as an action.
- **Utility**: The ability to run custom functions (based on Python) or APIs.
- **Filter**: We can filter results based on the previous block of results.

- **Decision**: We can make a decision and run different actions based on filtering.

- **Format**: This can be used to create body text that can be used when sending emails, creating tickets, and so on.

- **Prompt**: This is a request for user input (and this will be available for a user to select in the **Approvals** tab of the case or event).

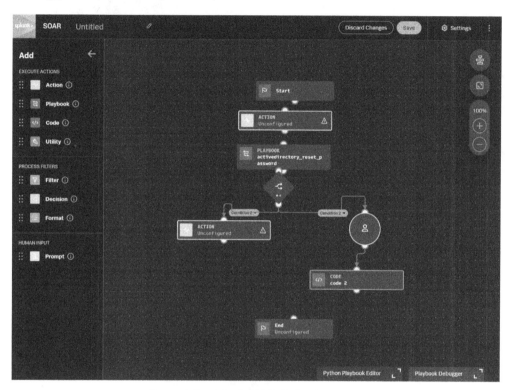

Figure 4.23 – Splunk SOAR playbook creation

For each playbook, you can also access audit and run statistics to evaluate whether you need to perform any changes.

Playbooks can be run on events and cases from their queue page, and you also have the option to run them on specific artifacts, all artifacts, or only new artifacts. As mentioned, we can run playbooks on artifacts from the **Artifacts** page.

If playbooks utilize an action to prompt a user for feedback, those prompts can be seen under the **Tasks** view or under **Approvals** for specific events/cases. Prompt actions can be configured with SLAs, so if an analyst doesn't respond, the playbook can fail or continue with a default value. Users also have the option to delegate a specific task to a different user or a role, which is vital if a user doesn't have enough permissions at that point to perform an action.

Figure 4.24 – The Splunk SOAR case approval view

As well as for playbooks, Splunk SOAR utilizes actions for automation. These actions are connected to the connector we add, and we can utilize these actions to respond quickly to certain situations. That can be an action to block a specific user or IP, add comments in an external SIEM solution, and so on. We can also schedule to run these actions in a specific event/case.

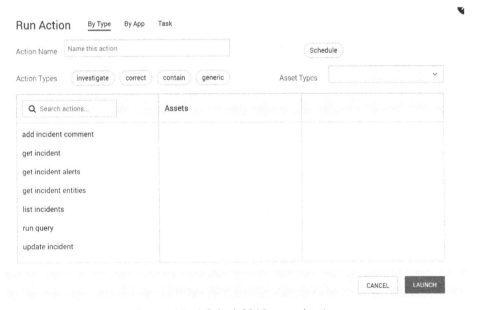

Figure 4.25 – A Splunk SOAR manual action

Next up, we will understand the reporting capabilities of Splunk SOAR.

Reporting

Splunk SOAR reporting is often used for creating reports that executives can refer to find out information about incident management and playbook utilization as well as the details about a specific case or event.

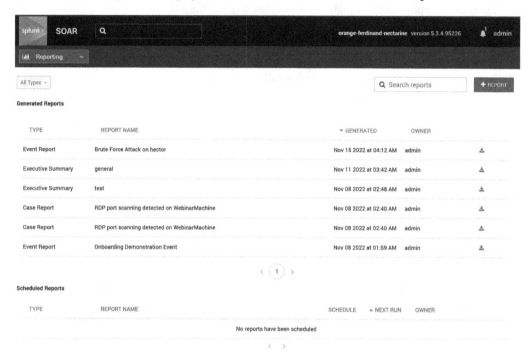

Figure 4.26 – Splunk SOAR reporting

This is an excellent tool for generating reports for management. Splunk SIEM has many more ways to generate reports utilizing data and events in SIEM itself. Splunk has an app that can send events from Splunk SOAR to SIEM to utilize that data for reporting (if you are ingesting events from third-party sources directly to Splunk SOAR and not Splunk SIEM).

TI and TVM

Splunk utilizes its SIEM solution to handle TI and TVM, as you can ingest data gathered from TI and TVM and work with that data in the SIEM solution itself. As TI and TVM are not integrated parts of Splunk SOAR, they will not be covered here. You can still utilize automation to enrich incidents with TI and TVM data, such as **Splunk Intelligence Management (TruSTAR) Indicator Enrichment**. More about this integration can be found at this URL: `https://www.splunk.com/en_us/blog/security/trustar-enrich-indicators-soar-in-seconds.html`.

The administration pane

Splunk SOAR allows you to configure quite a few things related to SOAR using its **Administration Settings** space. This is important as it will give SOCs the option to configure features such as access, which can then be applied to SOC analysts. We will list some of the most important ones here:

- Tags that you will use while managing events and cases.
- Workbooks for specific cases that we want to utilize in our investigation. These workbooks can have a few phases, and each phase can have different tasks assigned.
- Creating custom case statuses.
- Creating custom severity.
- An SLA for different severities.
- Roles and permissions.
- Two-factor authentication.
- Audit trail.
- Playbook run history.
- Action run history.

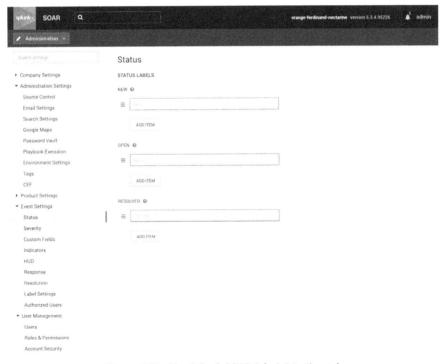

Figure 4.27 – The Splunk SOAR Administration tab

Thus far, we have introduced two strong SOAR tools – Microsoft Sentinel SOAR and Splunk SOAR. We saw what features they have and what options SOC analysts can use in their day-to-day tasks. Now, we will introduce the third and final SOAR tool – Google Chronicle SOAR.

Google Chronicle SOAR (Siemplify)

Google Chronicle is a well-known data analytics platform that is hosted on GCP. As the SIEM space evolved into the cloud, Google's security division joined and expanded into it. As with any other SIEM solution, it can collect data from any source – in local infrastructure or the cloud.

Google Chronicle utilizes detections that vary from context built using YARA-L rules to purpose-built detections and correlation with IoCs. Google Chronicle SOAR is a cloud solution and is now integrated into a Google Chronicle offering. Like Splunk, Google decided to buy a SOAR solution and integrate it into its offering, which is why you will often hear people referring to Google Chronicle SOAR as Siemplify. As Siemplify was only a SOAR solution, Google Chronicle SOAR has the ability to integrate with third-party SIEMs and ingest alerts, incidents, and events.

To connect Google Chronicle SOAR to any solution we want to use for incident management and automation, we need to utilize Marketplace. In it, we can find not only integrations with many different systems but also sample use cases that you can deploy into your environment.

Let's start by taking a look at the incident management capabilities of Google Chronicle SOAR.

Incident management

When you open Google Chronicle SOAR, the first page, called **HOMEPAGE**, will show you precisely what a SOC analyst needs:

- **My Cases**: This is a list of cases you can filter by those assigned to you as a SOC analyst (for example, John Stuart) or to your role (for example, Tier 1 Analysts), or whether it mentions you or your role in the incident chat. From here, we can see important details about a case, such as its name, severity, and how long it is active, as well as details about the case and links to it and whether we need to sort it by time created, priority, and SLA.

- **Pending Actions**: This shows whether there is any action waiting for my approval. We can filter these actions by severity as well.

- **My Tasks**: This has any task assigned to me or my role or created by me. In this space, we can create new tasks as well. We can filter them by status, sort them by time, or search for them. When creating a new task, we can assign tasks and pick up the due date in addition to the task title and description. When we have a task assigned to ourselves, we can view it or mark it as done, but if we created a task, we can also edit or delete it.

- **Requests**: Any open request is in this space, and we can create new requests as well.

- **Workspace**: This is a space where we can store important links, files, contacts, and notes, which can easily be found when needed. This is a great one-stop shop for all the important information we need in our day-to-day SOC life.

- **Announcements**: Here, we can see any announcements made or make an announcement. This announcement must have a title and description, and we can also configure an expiration date for it.

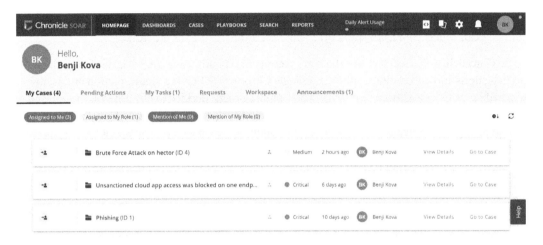

Figure 4.28 – The Google Chronicle SOAR home page

One more important tab in incident management is the **CASES** tab, where we can see a list of all cases in our environment including its details. Here, we can do the following:

- Sort cases by time created and modified, SLA, case stage, and so on.

- Filter cases by alert names, analysts, environments, priorities, products, stages, or tags. We also have the option to save filters for later use.

- Search for a specific case, refresh a view, or switch between the default and compact view.

- Select multiple cases, giving us the option to mark several/all cases and close or merge them into one case.

- Manually create a case. In this section, we can enter all the details of the case, such as title, creation reason, priority, assignment, alert details, entities, tags, and also the playbooks that will be run on an alert.

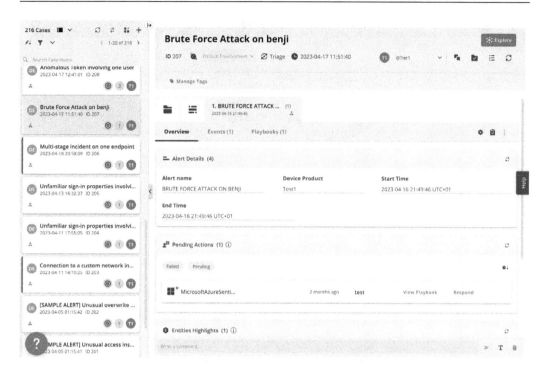

Figure 4.29 – The CASES view in Google Chronicle SOAR

We can utilize the **DASHBOARDS** tab if you need to see more complete pictures of your SOAR. It contains default views (**SOC Status** and **Playbook dashboards**) that you can edit, but you can also create your own views. When editing existing views or creating your own, you can utilize widgets. These widgets include pie charts, horizontal bar charts, vertical bar charts, tables, editor, image, ROI charts, and SOC status.

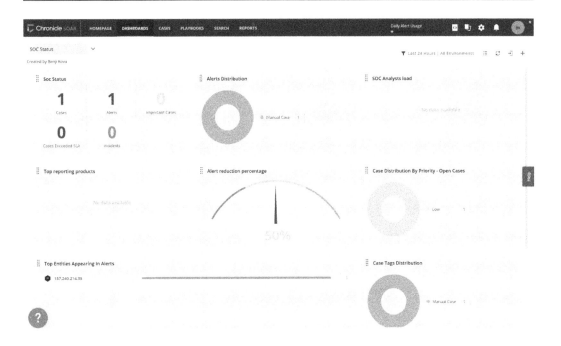

Figure 4.30 – The Google Chronicle SOAR dashboard

If we need to perform a more advanced search, we can utilize the **SEARCH** tab, which provides us with an option to search for cases or entities, for which we can utilize any filter associated with that case/entity.

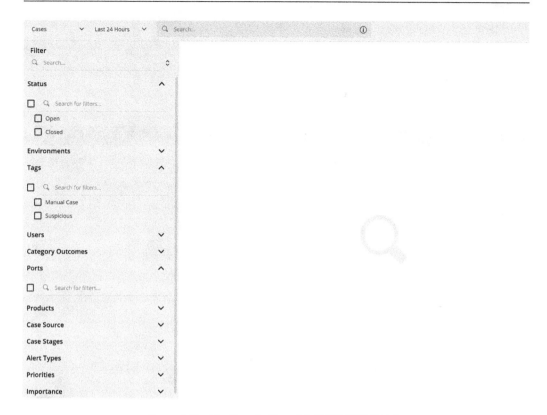

Figure 4.31 – The Google Chronicle SOAR SEARCH tab

The next section will familiarize you with the features available in Google Chronicle SOAR when it comes to incident investigation.

Investigation

An investigation can start from the **HOMEPAGE** or **CASES** tab by checking the details of each case. Some initial information that we can see is the case title, when the case was modified, what stage it is at, to whom it is assigned, and a list of alerts that we can investigate.

We can perform specific actions on a case, such as the following:

- Adding tags
- Assigning/reassigning the case
- Commenting/chatting with others regarding the case
- Closing the case
- Marking the case as important

- Marking the case as the incident

- Changing the case stage

- Changing the case priority

- Creating a case report

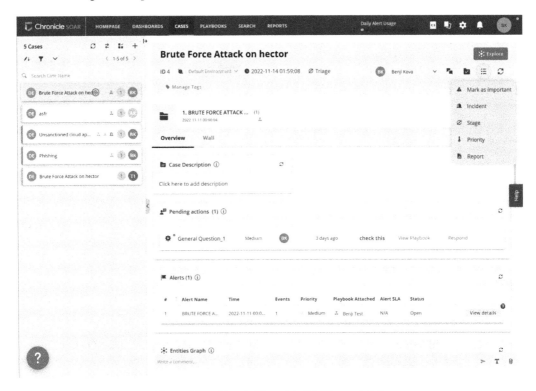

Figure 4.32 – Google Chronicle SOAR case details

The overall case will show us a case description, any pending actions, a list of alerts, an entity graph and highlights, the latest activity, statistics, and similar cases to the one we are investigating now. We also have a **Wall** section that will show us all activities on a case, allow us to filter between different activities, or show activities performed on specific alerts or by a specific user. We can mark any activity as a favorite for easier filtering later.

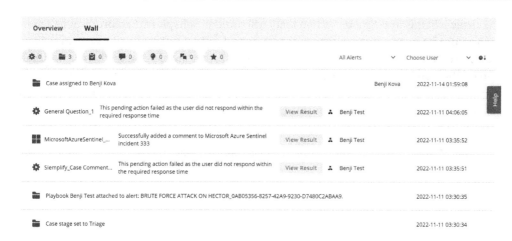

Figure 4.33 – The Google Chronicle SOAR case wall

Each alert will contain alert details, any pending action for this specific alert, highlighted entities, and events. The **Events** tab will contain more info about events detected, and we can view more details about an event or access event configuration (visualization and mapping). The last tab in this view is **Playbooks**, where we can see a playbook's runs, check its details, or rerun it if needed. We can also add a playbook to an alert as well as jump to the case wall to check activities done by playbooks.

What we can further do for a specific alert is to ingest it as a test case, move the alert to a different case, change the priority, add an entity, or close the alert itself.

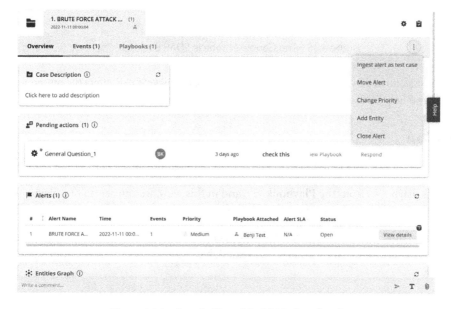

Figure 4.34 – Google Chronicle SOAR alert details

The final element in this *Investigation* section that I will mention is **Explore**. **Explore** gives us a more visual experience for cases, where we can see case alerts and entities assigned to the same case. From here, we can see alert and entity details, as well as play alerts as they are happening (there's a dedicated button to do this), and also the connection between them and entities. This can be helpful to SOC analysts to see how alerts were detected one by one and how they are connected. For each entity selected, we can add an entity property or run a manual action on each entity.

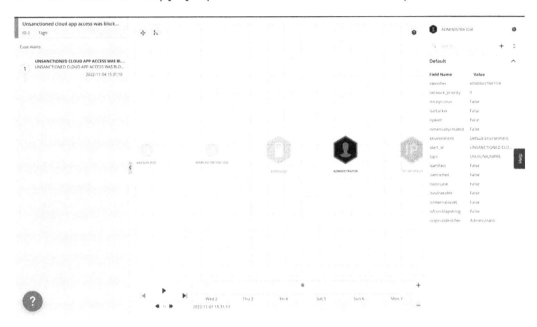

Figure 4.35 – The Google Chronicle SOAR Explorer

As in Microsoft Sentinel SOAR and Splunk SOAR, Google Chronicle SOAR offers the option to create automations and help SOC analysts perform faster investigations and responses.

Automation

Google Chronicle SOAR uses two different methods for automation – *playbooks* and *manual actions*.

We can access playbooks from the **Playbook** tab, and in this view, we can see all our playbooks. We can arrange playbooks by folders, export them, import playbooks, and of course, create new ones.

When we are creating playbooks, we start with **Trigger**. The trigger will tell a playbook when it will run, and this trigger can be when an alert is created, based on an alert type, custom list, custom trigger, and so on.

Once we have a trigger, we need to add **Actions**. Actions will do what we need once a playbook is triggered. This can be enrichment, response, or orchestration. For example, when an alert is created and we have a user, we want to get all the groups that the user is a member of from Azure AD to check whether the user is part of some sensitive group. Actions are part of the connections we add to integrate solutions with SOAR – in this case, the Azure AD solution.

We also have **Flows** as a segment of the playbook, and these flows can be conditions, multi-choice questions, or conditions based on previous actions.

Google Chronicle SOAR has different views in playbooks, where we can create different views for SOC analysts so that they can quickly find data collected with the playbook. These views are customizable using widgets.

Another great tool in playbooks is a playbook simulator, which can help you debug your playbook in real time.

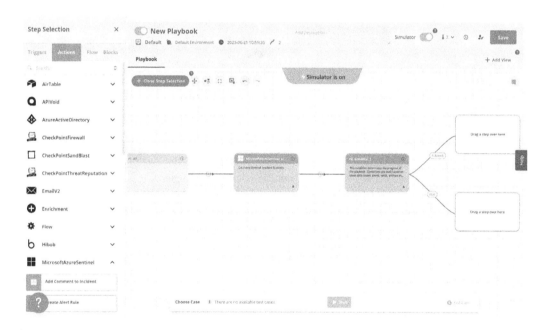

Figure 4.36 – Google Chronicle SOAR playbook creation

We can assign these playbooks to specific alerts and see the run of each of these playbooks in our cases.

If we use playbooks that interact with SOC analysts, a SOC analyst would need to provide feedback before a playbook continues to run, and then we can see all the pending actions from those running playbooks in **Pending actions** on **HOMEPAGE**.

Google Chronicle SOAR has support for manual actions as well. Manual actions, like actions in playbooks, will depend on integrations we perform (Azure AD, firewall systems, external SIEMs, and so on).

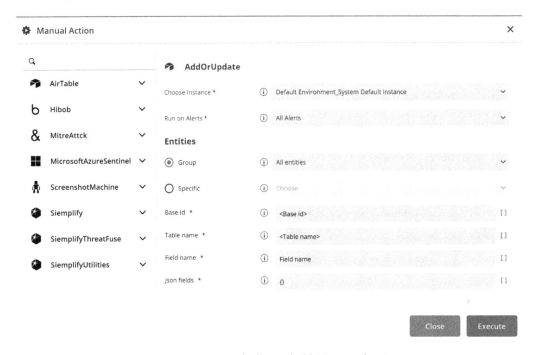

Figure 4.37 – A Google Chronicle SOAR manual action

We can run these manual actions on any case, and we can choose whether we want to run them on all alerts or specific ones, all entities or specific ones, and also add specifics about actions we run. This provides a quick and easy way to perform any action we need, such as blocking specific IPs in a firewall or blocking users in Azure AD.

Reporting

Reporting in Google Chronicle SOAR gives us different report categories that we can choose from if we want to deliver a report to management, one that will contain ROI data, SLA data, a general report, and so on. We can edit these reports by adding an editor, pie charts, a table, and a vertical bar chart, and we can also schedule report creation.

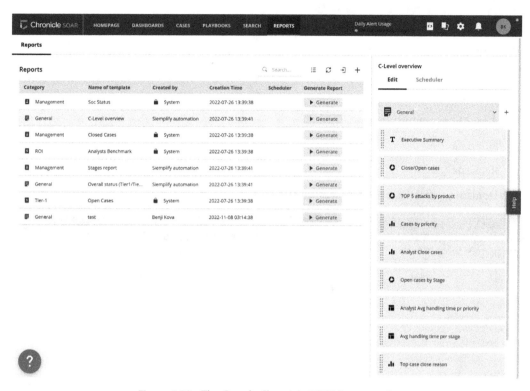

Figure 4.38 – The Google Chronicle SOAR Reports tab

For more detailed information about raw events, we can use Google Chronicle SIEM or any other connected SIEM source, as Google Chronicle has separate SIEM and SOAR environments.

TI and TVM

Google Chronicle utilizes its SIEM solution to handle TI and TVM, as you can ingest them and work with that data in the SIEM itself. As TI and TVM are not integrated parts of Google Chronicle SOAR, they will not be covered here. You can still utilize automation to enrich incidents with TI and TVM data.

Administration pane

Google Chronicle SOAR allows you to configure quite a few things related to SOAR. We will list some of the most important ones:

- User management, permissions, and roles
- Tags management
- Case stages

- The root cause of a closed case

- A case name

- Alert grouping settings

- Audit user activities

- A multi-environment policy

- SLA data

- Requests

- Domains, networks, a blocklist, and so on

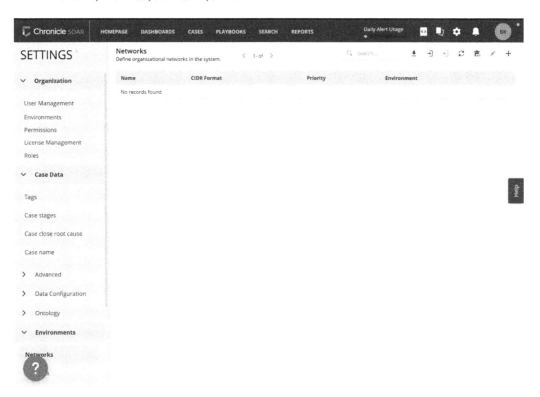

Figure 4.39 – The Google Chronicle SOAR Administration tab

With that, we have covered the main Google Chronicle SOAR functionalities when it comes to incident management, investigation, automation, and reporting.

Summary

In this chapter, we introduced three important SOAR tools and guided you through the SOAR elements of the tools themselves. We introduced Microsoft Sentinel SOAR, hosted in Microsoft Azure, which is fully integrated into Microsoft Sentinel SIEM. Microsoft Sentinel is a cloud-native solution, and as such, there is no version for local infrastructure.

The second tool was Splunk SOAR, and we used a local infrastructure version of the solution, although it is also available as a cloud option. The third solution, and again, only available in the cloud, is Google Chronicle SOAR, which is hosted on GCP.

While there are a few other popular solutions, such as Palo Alto SOAR, most of the functionalities are similar to the ones already covered. This should not remove the need to explore even deeper on your own.

The next chapter will dive deep into Microsoft Sentinel automation, which we demonstrate use cases for in *Chapters 6 to 8*.

5
Introducing Microsoft Sentinel Automation

In the previous chapter, we introduced a few SOAR tools and some of the main features we can utilize in our day-to-day operations. We showcased what incident management, investigation, automation, and reporting look like in real tools and offered some directions on how to utilize them.

This chapter will focus on Microsoft Sentinel automation, and we will dive deep into each element when working with it. We will discuss automation rules, playbooks, their elements and permissions, and prepare you for hands-on examples that will be covered in *Chapters 6 to 8*.

In this chapter, we will discuss the following:

- The purpose of Microsoft Sentinel automation
- All about automation rules
- All about playbooks
- Monitoring automation rules and playbook health

The purpose of Microsoft Sentinel automation

Microsoft Sentinel automation's purpose, like the purpose of all automation, is to take repetitive tasks and transform them into automated tasks. In the SOC, some of the topics automation focuses on are as follows:

- **Enrichment**: When an incident is created, we want to enrich it with additional data. This will save SOC analysts time as they will have this enrichment as soon as they pick up the incident. For example, when an incident with an IP address is created, we can run an automatic playbook to enrich the incident with TI data about whether the IP is known to be malicious or not.

- **Initial triage and incident suppression**: This accompanies enrichment as we can utilize the results of that to decide whether we want to auto-close an incident if the IP is internal and behavior is expected, or transfer it to tier 2 if the IP address is known to be malicious.

- **Orchestration**: This is more oriented to orchestrating incident assignments or notifying SOC analysts that an incident has been created or assigned to them. For example, we can create automation that will send a Microsoft Teams chat message to the user using adaptive cards on incident creation, from where we can utilize SOC analyst input to even auto-close incidents if the information that's shared warrants making that decision.

- **Response**: SOC analysts used to struggle when they needed to block an IP address, isolate a machine, block a user or reset their password, and so on. Because SOC analysts wouldn't usually have permission to access a firewall, active directory, or EDR solution, they would need to raise an internal ticket or ping the network or system administrator to help out. By running a playbook, that task can be done for them automatically. This is crucial for many modern threats, where the time it takes to contain the threat must be minimal.

To perform any of the preceding tasks, Microsoft Sentinel uses two different automation methods:

- **Automation rules**: These are used to manage automation in Microsoft Sentinel centrally. Automation rules contains triggers, conditions, and actions that dictate how an automation rule will respond. In the next section, *All about automation rules*, we will dive deep into this feature.

- **Playbooks**: Playbooks are a list of actions that will be performed on an incident. This can include enrichment, response, remediation, and much more. We will cover playbooks in more detail later, in the *All about playbooks* section.

All about automation rules

As mentioned previously, automation rules can be created to manage automation in Microsoft Sentinel centrally. But how can we do that?

Automation rules in Microsoft Sentinel have three main aspects:

- Triggers

- Conditions

- Actions

Automation rules are sorted in order, which is a critical element since all automation rules will run from the lowest order number (for example, 1) to the highest (for example, 55), and they will run sequentially.

However, before we go into more detail about triggers, conditions, and actions, let's familiarize ourselves with the **graphical user interface** (**GUI**) of Microsoft Sentinel automation rules and learn more about permissions.

Navigating the automation rule GUI

Microsoft Sentinel automation rules are located under the **Automation** tab in the **Automation rules** sub-menu. In this menu, we have the option to create an automation rule, edit an automation rule, enable or disable an automation rule, move it up or down, remove an automation rule, as well as filter automation rules by analytic rules, actions, triggers, statuses, who created them, and when they were last modified.

Figure 5.1 – Microsoft Sentinel automation rules

We need to click on **Create** and select **Automation rule** to create an automation rule.

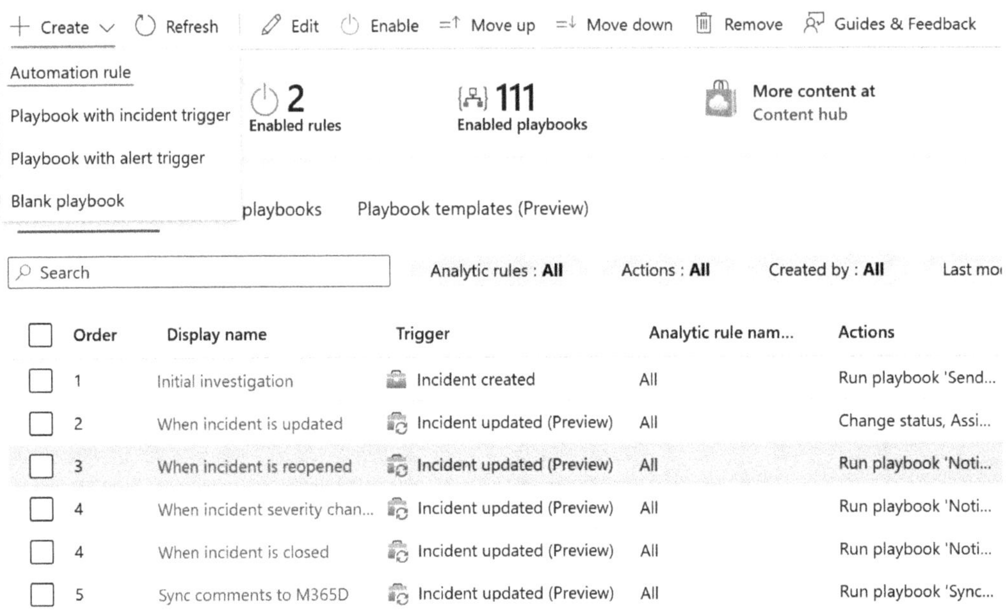

Figure 5.2 – Creating a new automation rule

From here, we must navigate to the **Create new automation rule** window; this is where we can start creating this new rule.

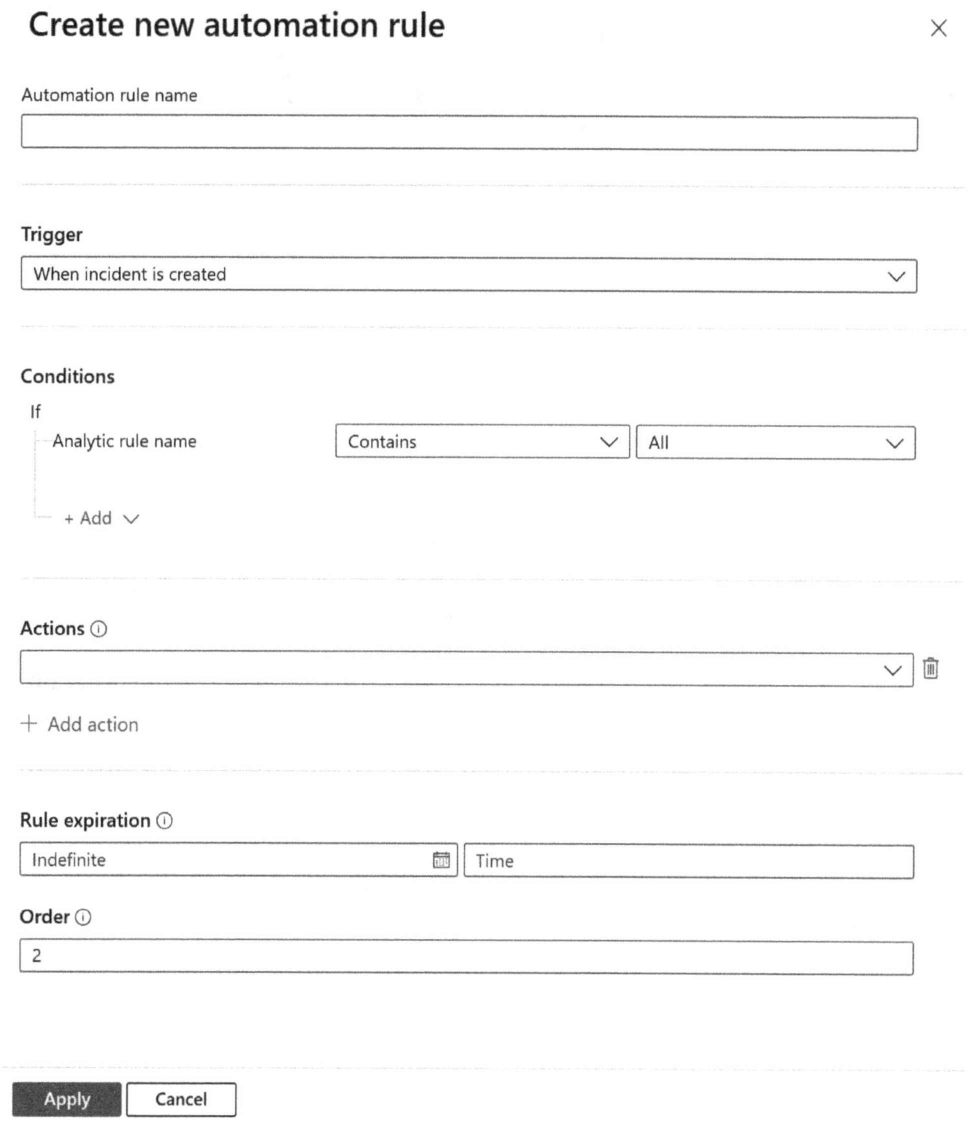

Create new automation rule   ✕

Automation rule name

[   ]

Trigger

| When incident is created   ⌄ |

Conditions

If

├─ Analytic rule name | Contains   ⌄ | | All   ⌄ |

└─ + Add ⌄

Actions ⓘ

[⌄] 🗑

+ Add action

Rule expiration ⓘ

| Indefinite   📅 | | Time |

Order ⓘ

| 2 |

[**Apply**] [Cancel]

Figure 5.3 – The Create new automation rule wizard

However, there are other ways to create or edit automation rules:

- When creating an analytic rule using the **Analytics rule wizard** area, under the **Automated response** tab, we can see what automation rules will be triggered when an incident is created. We can also create a new automation rule specific to this rule. We can choose between any trigger at this stage.

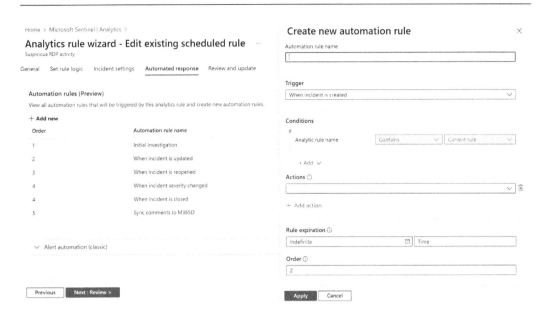

Figure 5.4 – Creating an automation rule when creating the analytic rule

- From the **Incidents** page, we can select an incident and, from the right menu, under **Actions**, select **Create automation rule**.

Figure 5.5 – Creating an automation rule from the Incidents page

Using this method, automation rule conditions will be filled with the data that was detected in the incident itself.

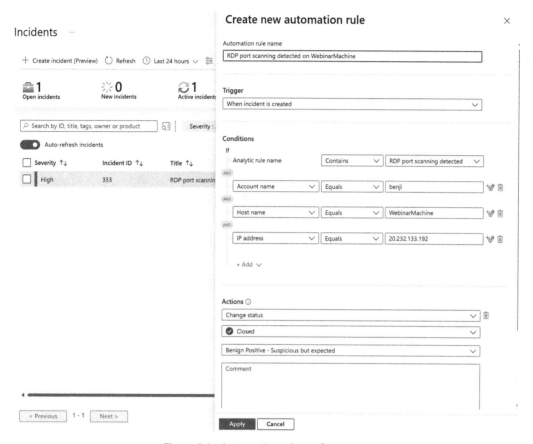

Figure 5.6 – Automation rule configuration

- We can follow the same steps to create an automation rule from the **Incident investigation** page.

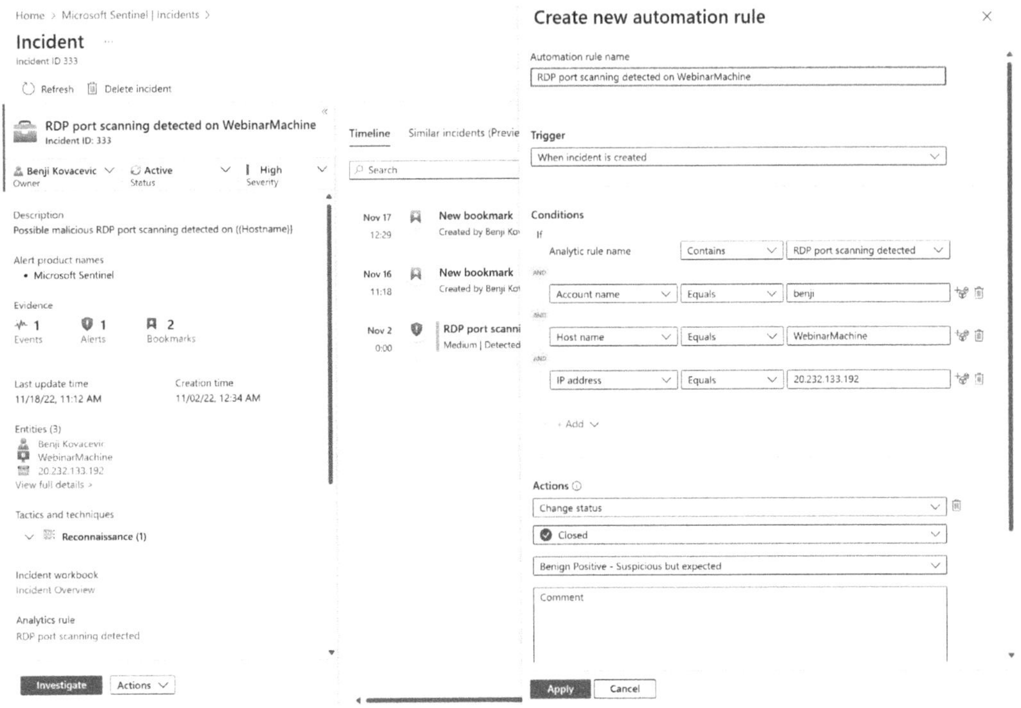

Figure 5.7 – Creating an automation rule from the Incident Overview page

To be able to create automation rules or playbooks, users must have the right permissions to create or edit them. Let's go through the permissions for Microsoft Sentinel automation.

Permissions

To create automation rules, a user needs to have a **Microsoft Sentinel Responder** or **Microsoft Sentinel Contributor** role assigned.

There is one more special role connected to Microsoft Sentinel automation rules – **Microsoft Sentinel Automation Contributor**. This is not a user role but instead a role that needs to be assigned to a Microsoft Sentinel identity so that an automation rule can run a playbook as an action. This is assigned to Microsoft Azure resource groups, which is where playbooks reside. For example, if we have five resource groups that contain playbooks, we want to have the option to attach them as an action; we need to assign this permission to all five Microsoft Azure resource groups.

To assign this permission, we need to go to the Microsoft Sentinel instance, go to **Settings**, then **Settings** again, and then, under **Playbook permissions**, click on **Configure permissions**. In the next window, choose the resource groups you want to assign the **Microsoft Sentinel Automation Contributor** role.

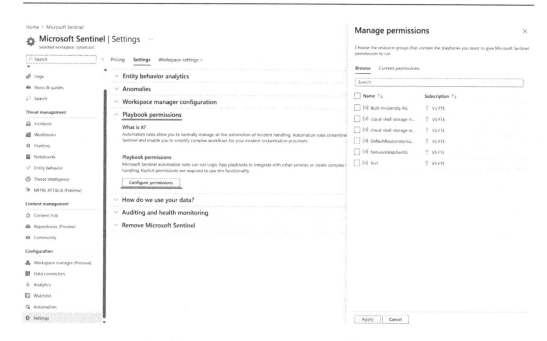

Figure 5.8 – Microsoft Sentinel playbook permissions configuration

Once the permission is applied on the resource group level, we can attach playbooks from that resource group to an automation rule as an action.

Triggers

Triggers are used to define when an automation rule runs. For automation rules, we have three different triggers:

- **When incident is created**: This supports a complete list of conditions
- **When incident is updated (Preview)**: This supports a complete list of conditions, plus conditions associated with information about updated data
- **When alert is created (Preview)**: This supports only one condition (an analytic rule name)

Figure 5.9 – Automation rule triggers

Conditions

Once a trigger has been set, we need to set conditions. Conditions are used to filter what incidents we want to run specific actions on, as we don't want the same actions on all incidents. What's important to note here is that actions will run only if all conditions are met. Automation rules support **OR** and **AND** condition grouping, which allows us to create more detailed automation.

An analytic rule name is one condition that cannot be removed and is used across all triggers. The evaluation supports **Contains** and **Does not contain** options, while for values, we can choose all analytic rules or run on only specific analytic rules created in Microsoft Sentinel. When we choose **All** as a value for incident creation and update, this will also run on synchronized incidents from tools such as Microsoft 365 Defender and Microsoft Defender for Cloud.

Since we have multiple triggers in automation rules, let's see what conditions are supported for each.

Conditions associated with the "When incident is created" trigger

The **When incident is created** trigger can only check the current state of the values of an incident. If we evaluate the same incident in multiple automation rules, the current state can change if we update the incident in a previous automation rule. For example, if the severity of incident creation is set to **Medium**, that will be the current state for the first automation rule. Suppose, in the automation rule, we take action to change the severity to **High**. In that case, the current value for severity in the following automation rule (that is, automation rule number 3) that will run on this same incident will be **High**.

The following conditions can be used with the **When an incident is created** trigger:

- **Incident properties**: For example, analytic rule name, title, description, severity, owner, status, tactics, and custom details

- **Entity properties**: For example, account name, account domain, filename, file hash, hostname, IP address, IoT device, mail message details, URL, and many more

Figure 5.10 – Automation rule condition values

Based on the condition selected, we can use one of the following evaluation methods:

- **Equals** or **Does not equal**
- **Contains** or **Does not contain**
- **Starts with** or **Does not start with**
- **Ends with** or **Does not end with**

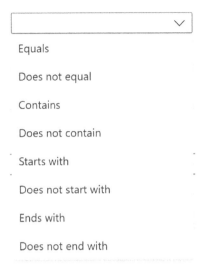

Figure 5.11 – Automation rule condition options

The final element of the condition is the value itself. Here, we can have clear text input (such as an incident title or description) or the option to select pre-existing values (such as incident severity or status). We can also add multiple inputs to the same condition.

Figure 5.12 – An automation rule condition example

Conditions associated with the "When incident is updated" trigger

If we update an incident with one of the supported conditions, we can trigger automation rules based on that. This gives us complete control over automation scenarios to support incident creation and updates.

Conditions with an incident update trigger support **current state** and **state changes**. Conditions include those from the incident creation trigger, plus the following:

- **Owner**
- **Updated by**
- **Alerts**
- **Comments**

They also include **state change** values for the following:

- **Severity**
- **Status**
- **Tactics**
- **Tag**

The following evaluation methods can be added with **state change**:

- **Changed** (owner, severity, and status)
- **Changed To** (severity and status)
- **Changed From** (severity and status)
- **Added** (alerts, comments, tactics, and tags)

The only difference is **Updated by**, which uses the current state for evaluation. We can select one of the following values from the dropdown:

- **Application**
- **User** (a manual change of a field by a specific user)
- **Alert grouping** (adding an alert to the incident)
- **Playbook** (a change made by a playbook run)
- **Automation rule** (a change made by an automation rule run)
- **Microsoft 365 Defender** (for updates to incidents made by bidirectional incident synchronization from Microsoft 365 Defender)

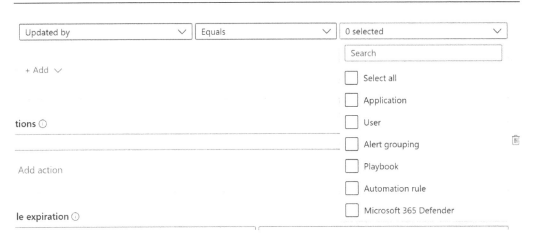

Figure 5.13 – The Updated by automation rule values

Conditions associated with the "When alert is created" trigger

The **When alert is created** trigger only supports analytic rules created in Microsoft Sentinel, and the only supported condition is **Analytic rule name**.

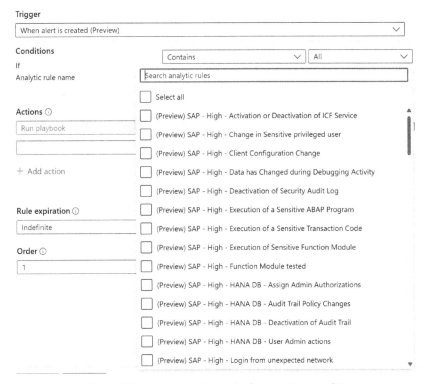

Figure 5.14 – An automation rule alert creation condition

After choosing one of the triggers (incident creation, incident update, or alert creation) and configuring the conditions, if all conditions we configured are met, we can run one or more actions on an incident.

Actions

The following actions are supported for automation rules:

- **Run playbook**:

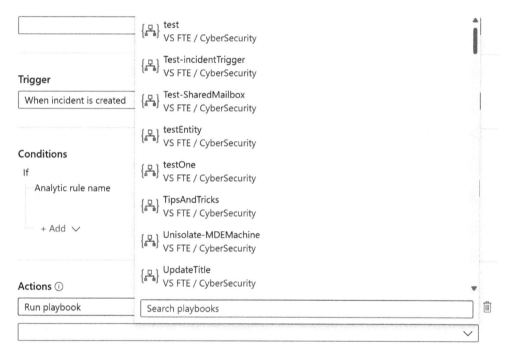

Figure 5.15 – Run playbook action

- **Change status**:

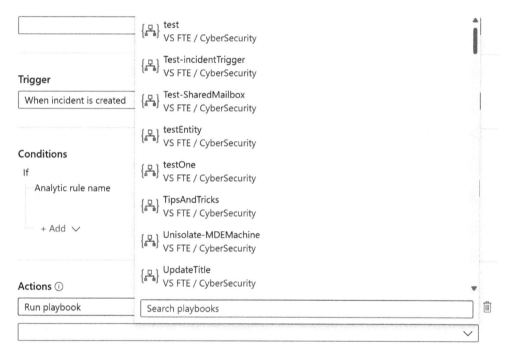

Figure 5.16 – Change status action

- **Change severity:**

Figure 5.17 – Change severity action

- **Assign owner:**

Figure 5.18 – Assign owner action

- **Add tags**:

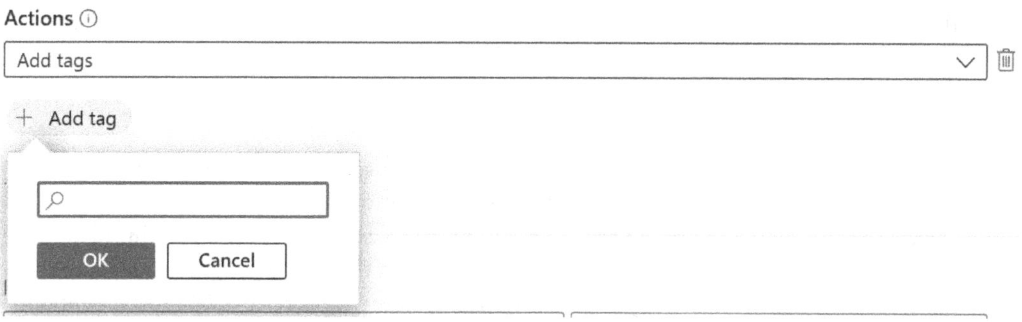

Actions ⓘ

| Add tags | ∨ | 🗑 |

+ Add tag

Figure 5.19 – Add tags action

> **Important note**
> The **Run playbook** action is the only action that's available when utilizing the **When alert is created** trigger.

While automation rule triggers, conditions, and actions are fields, we will always want to configure to have effective automation rules; we can also configure rule expiration and order.

Rule expiration and order

As mentioned previously, we have two additional steps we can configure when creating automation rules:

- **Rule expiration**: This is if we are creating an automation rule that will be active for only a specific period – for example, if we are performing penetration testing. We want to auto-close incidents created by and during penetration testing and disable the automation rule at a specific date and time.

- **Order**: This is what execution order number we want to configure an automation rule as. To recall, all automation rules are run via an order number, from a lower number to a higher one, sequentially.

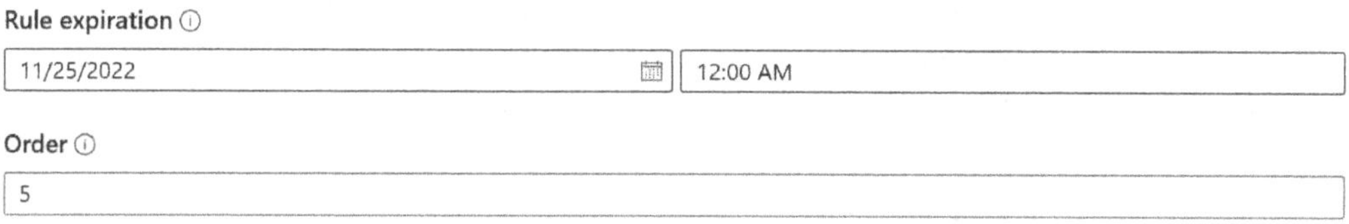

Rule expiration ⓘ

| 11/25/2022 | 📅 | 12:00 AM |

Order ⓘ

| 5 |

Figure 5.20 – Automation rule expiration and order configuration

With that, we have covered all the major elements of automation rules and how they work. In the next section, we will look at playbooks and their main building blocks in Microsoft Sentinel automation in more detail.

All about playbooks

Playbooks are a list of actions that will be performed on the incident. They can include enrichment, response, remediation, and much more. To achieve this, Microsoft Sentinel utilizes a Microsoft Azure solution called **Logic Apps** – a platform used to create and run automated workflows. This platform uses low- or no-code and focuses more on visual design. However, those who prefer to code more can utilize coding mode as well. Because of this, it is common to hear people refer to Microsoft Sentinel playbooks as Logic Apps.

There are two different types of Logic Apps that Microsoft Sentinel supports:

- **Logic Apps Consumption**: This is a single playbook that has only one workflow. It supports templates and custom connectors and is widely integrated into Microsoft Sentinel with template support. Logic Apps Consumption shares the same backend resources across different customer tenants. We will use the Logic Apps Consumption model in our hands-on examples.

- **Logic Apps Standard**: This is a single Logic App that can have multiple workflows. It doesn't support templates and custom connectors, which is why Microsoft Sentinel doesn't have playbook templates created in Logic Apps Standard. In Logic Apps Standard, workflows in the same Logic App share the same backend resources, and they are not shared across different Logic Apps like they are with Logic Apps Consumption. It's also important to note that when creating a Logic Apps Standard playbook, it must be stateful and cannot utilize private endpoints – Microsoft Sentinel does not support these scenarios at the time of writing.

Microsoft Sentinel is a unified way to run a playbook, and it will make no difference whether Logic Apps Consumption or Logic Apps Standard is used.

Navigating the playbooks GUI

Microsoft Sentinel playbooks are located under the **Automation** tab in the **Active playbooks** sub-menu. In this menu, we have the option to create a playbook, open playbook details to edit or manage it, enable or disable a playbook, delete a playbook, as well as to filter playbooks by status, trigger kind, subscription, resource group, plan, and source name. If we have deployed the playbook using built-in templates, we will also get information on whether an update is available.

Figure 5.21 – Microsoft Sentinel playbooks

Playbooks support templates; all deployed templates can be found in the **Playbooks templates (Preview)** sub-menu. We can deploy any playbook, from templates to active state. We can filter templates by trigger, Logic App connector, entities, tags, and source name. If we have already deployed a playbook template, we will see a notification stating that a specific playbook is in use.

Figure 5.22 – Microsoft Sentinel playbook templates

To access all templates in Microsoft Sentinel, we can utilize **Content hub** and the available solutions, where we can filter, among others, by the solution we need or solutions with playbook templates.

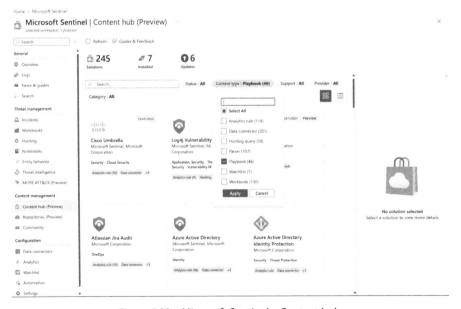

Figure 5.23 – Microsoft Sentinel – Content hub

More templates can be found on GitHub and can be easily deployed to Microsoft Sentinel since they utilize **Azure Resource Manager** (**ARM**) templates. Microsoft Sentinel has an official repository with lots of content available that is ready to be deployed. The link to the official repository is `https://github.com/Azure/Azure-Sentinel`.

To create a new playbook, go to the **Automation** tab, click **Create**, and select one of the following options:

- **Playbook with incident trigger**
- **Playbook with alert trigger**
- **Blank playbook**

Figure 5.24 – Creating a new Microsoft Sentinel playbook

If we select **Playbook with incident trigger** or **Playbook with alert trigger**, we will create a **Logic Apps Consumption** Logic App. The first view is where we enter basic information, such as what subscription and resource we want to deploy the playbook in, the region, and the playbook's name. We can also enable diagnostic settings, which we will cover in the *Monitoring automation rules and playbook health* section:

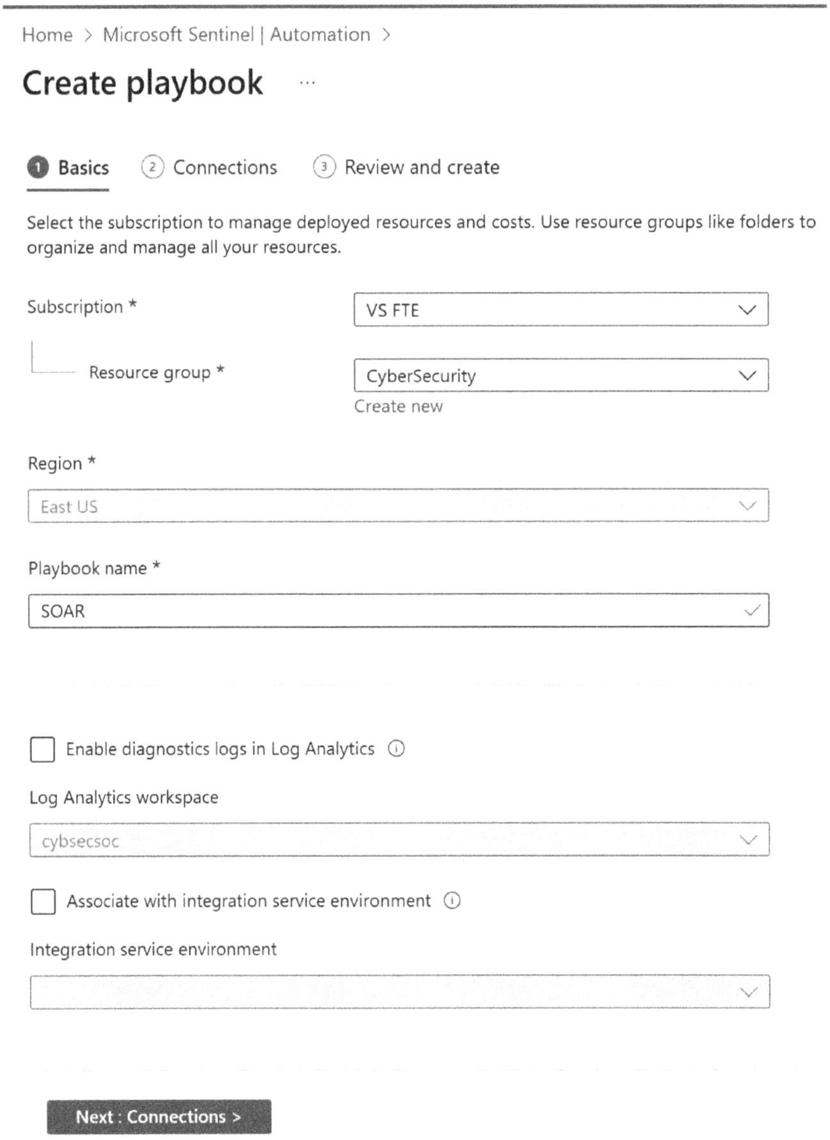

Figure 5.25 – The Create playbook wizard

In the next window, we can select how we want to authenticate a Microsoft Sentinel connection. By default, playbook creation will enable a **system-assigned managed identity** from the playbook and utilize it (the recommended method). However, we can utilize any pre-existing connection or change it once the playbook has been deployed.

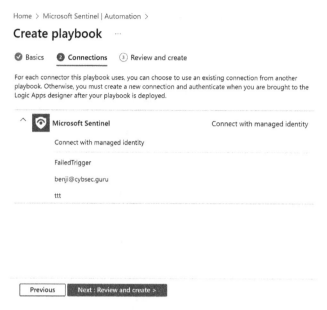

Figure 5.26 – Creating a new playbook – Connections

The last step is to review the configuration and create our playbook.

Figure 5.27 – Creating a new playbook – Review and create

Once the playbook has been deployed, we can navigate to **Logic app designer**, where we can start our playbook design.

Figure 5.28 – The Logic app designer view

From here, we can also access the code view of the playbook if we prefer to work with code.

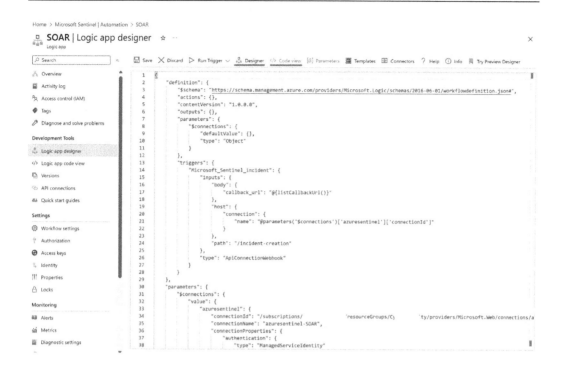

Figure 5.29 – The Logic App designer code view

We will go through the whole process of creation and explanation in *Chapters 6* to *8*, where we will cover hands-on examples.

When we want to create a blank playbook, we can choose between creating a **Logic Apps Standard** or **Logic Apps Consumption** Logic App. We can also utilize any other trigger available in Logic Apps, such as a recurrence to perform a regular playbook run, and a Microsoft Forms trigger to create an incident when a new form is filled in and when an email is received.

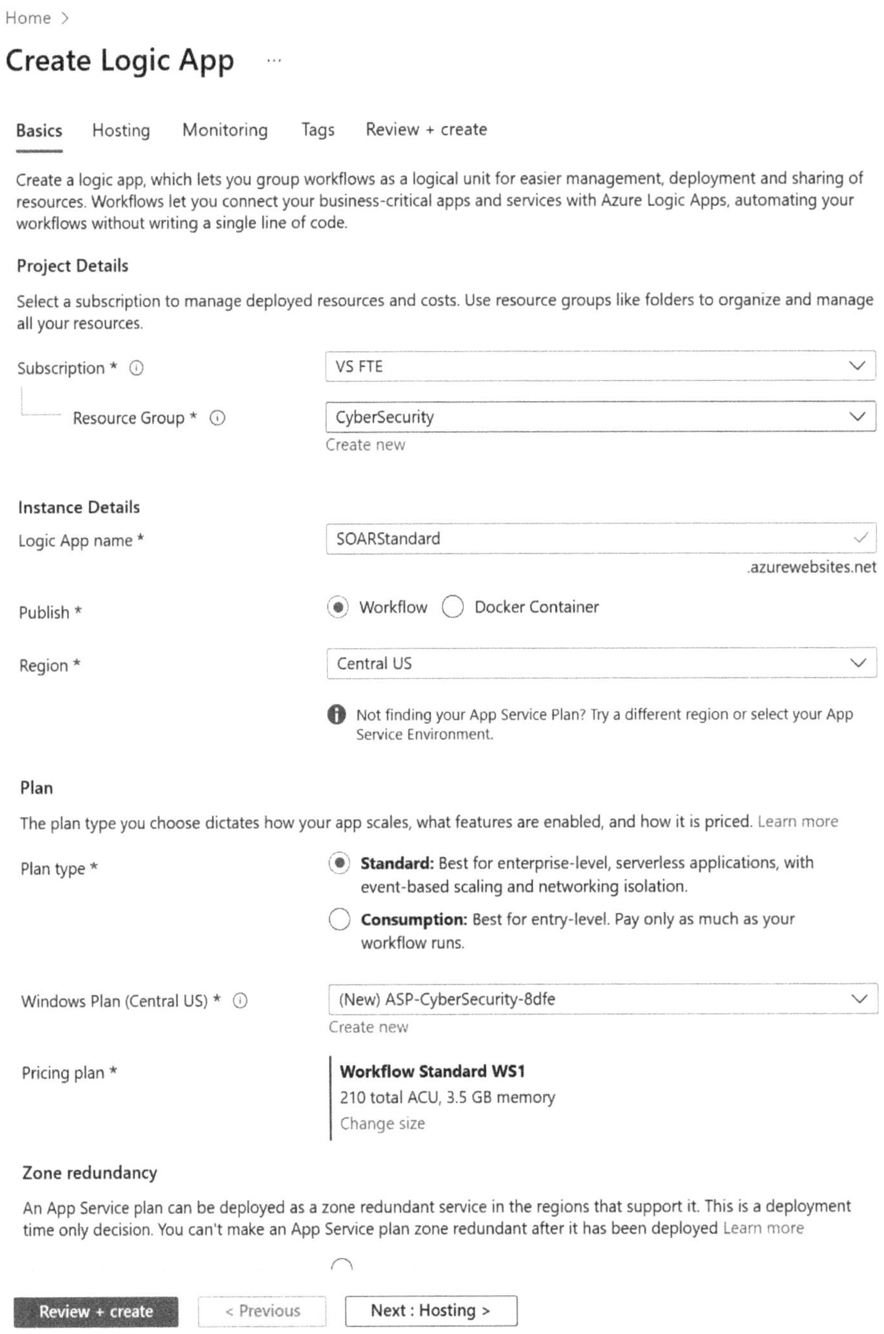

Home >

Create Logic App   ...

Basics   Hosting   Monitoring   Tags   Review + create

Create a logic app, which lets you group workflows as a logical unit for easier management, deployment and sharing of resources. Workflows let you connect your business-critical apps and services with Azure Logic Apps, automating your workflows without writing a single line of code.

Project Details

Select a subscription to manage deployed resources and costs. Use resource groups like folders to organize and manage all your resources.

Subscription * ⓘ | VS FTE   ∨ |

    Resource Group * ⓘ | CyberSecurity   ∨ |
                                        Create new

Instance Details

Logic App name * | SOARStandard   ✓ |
                                                               .azurewebsites.net

Publish *   ● Workflow   ◯ Docker Container

Region * | Central US   ∨ |

                                      ⓘ Not finding your App Service Plan? Try a different region or select your App
                                        Service Environment.

Plan

The plan type you choose dictates how your app scales, what features are enabled, and how it is priced. Learn more

Plan type *   ● **Standard:** Best for enterprise-level, serverless applications, with
                                         event-based scaling and networking isolation.

                                      ◯ **Consumption:** Best for entry-level. Pay only as much as your
                                         workflow runs.

Windows Plan (Central US) * ⓘ | (New) ASP-CyberSecurity-8dfe   ∨ |
                                        Create new

Pricing plan *   **Workflow Standard WS1**
                                      210 total ACU, 3.5 GB memory
                                      Change size

Zone redundancy

An App Service plan can be deployed as a zone redundant service in the regions that support it. This is a deployment time only decision. You can't make an App Service plan zone redundant after it has been deployed Learn more

[Review + create]   [< Previous]   [Next : Hosting >]

Figure 5.30 – Creating a playbook using Logic Apps Standard

To run the created playbook, we have a few options:

- Attach the playbook to an automation rule for automatic triggering (which will require the **Microsoft Sentinel Automation Contributor** role to be assigned to a Microsoft Sentinel identity; more about this in the subsequent *Permissions* section)

- Run the playbook manually on the incident (which will require the **Microsoft Sentinel Automation Contributor** role assigned to a Microsoft Sentinel identity)

- Run the playbook manually on the alert

To create, edit, and run playbooks in Microsoft Sentinel, you will need certain permissions to perform these actions. Let's go through the different permissions users can have and what users can perform with these actions.

Permissions

> **Important note**
>
> To understand this segment better, I suggest that you have a basic understanding of permissions on Azure and Azure RBAC. A great starting point is the official documentation: `https://learn.microsoft.com/azure/role-based-access-control/`.

There are a few different permissions that users can utilize based on the actions they need to perform when working with Microsoft Sentinel playbooks:

- **Logic Apps Contributor**: This gives you permission to manage Logic Apps and run playbooks, but you cannot change access to them (there is standard role separation in Azure, and only the **Owner** or **User Access Administrator** role can perform this action).

- **Logic App Operator**: This gives you permission to read, enable, or disable a playbook, but you cannot edit, update, or run playbooks.

- **Microsoft Sentinel Contributor**: This permits you to attach a playbook to an analytic rule, among other Microsoft Sentinel permissions.

- **Microsoft Sentinel Responder**: This permits you to run playbooks manually, among other Microsoft Sentinel permissions.

- **Microsoft Sentinel Playbook Operator**: This permits you to list and run playbooks manually.

We covered the **Microsoft Sentinel Automation Contributor** role earlier in this chapter, so we will not look at it in detail again.

To be able to create and utilize playbooks, it is important to understand how they work. In the next section, we will cover the main aspects of Logic Apps.

Logic Apps connectors and authentication

Under the hood, Logic Apps uses API calls to connect with Microsoft and non-Microsoft solutions. Those API calls can be wrapped in a Logic Apps connector, giving us more straightforward configuration and authentication. These connectors are as follows:

- **Managed connectors**: These are available in a Logic App out of the box. They contain triggers and actions for specific products or services, such as the Microsoft Sentinel connector. There are hundreds of managed connectors for Microsoft products and services, as well as for non-Microsoft products and services.

- **Custom connectors**: If some product or service still doesn't have a built-in connector in Logic Apps, you have the option to create and utilize a custom connector in your environment. It is also possible to share those custom connectors with others, and some of them are utilized with Microsoft Sentinel. When using custom connectors, you must utilize the Logic App Consumption model since Logic App Standard doesn't support custom connectors at the time of writing.

But what if there is neither a built-in nor custom connector?

In this case, we can utilize an HTTP connector that will allow us to connect to a product or solution using direct API calls. Examples of these HTTP calls will be covered in *Chapter 9*.

> **Important note**
>
> Data connectors in Microsoft Sentinel aren't the same as Logic Apps connectors.
>
> Data connectors in Microsoft Sentinel are used to ingest logs into a Log Analytics workspace, and we can utilize those logs to create detection rules, hunt for data, and so on. Some examples of these logs include Syslog data, security event data from Windows Server, and sign-in logs from Azure AD.
>
> Logic Apps connectors are API calls to products and services so that we can perform specific actions. Examples of these API calls are a call to Azure AD to block a user, a call to an EDR solution to isolate the machine, and an API call to the TI solution to get IP address information.

But wait! When making an API call, don't we need to provide authentication? Is this supported in Microsoft Sentinel playbooks?

Yes! If we need to authenticate managed or custom connectors, there is a way to do this. For non-Microsoft products and services, these can be usernames and passwords, API tokens, and so on. These authentications are saved as API connections and can be accessed from a playbook. Once created, these API connections can also be utilized in other playbooks; they are not specific to one playbook.

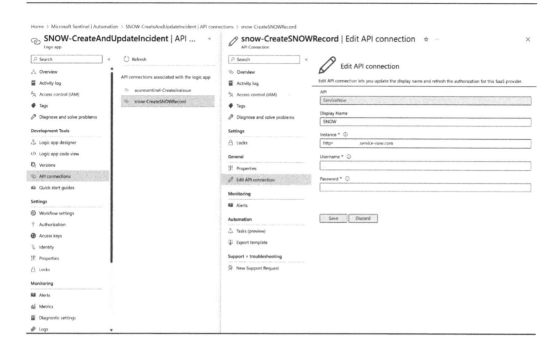

Figure 5.31 – Managing an API connection in a playbook

If we use HTTP calls, this information will be inserted into the header or body of the API call, as per the instructions of the product or service.

In terms of Microsoft services and products, playbooks support three types of API authentication. Let's look at them in detail.

System-assigned managed identity

This is the preferred option and is utilized by default when creating playbooks using **Create playbook with an incident trigger** or **Create playbook with an alert trigger**. Each playbook has its own system-assigned managed identity that can be enabled, and this identity can be utilized only by this specific playbook. This connection cannot be shared among playbooks. A managed identity also provides an option for the least privileged approach.

It's important to note that not all connectors in Logic Apps support managed identities. For example, a Microsoft Sentinel connector supports them, while Microsoft Teams and Office 365 Outlook do not.

To add permissions to a managed identity, you will need to go to the **Identity** tab in Logic Apps.

Figure 5.32 – Enabling a playbook's managed identity

Once you are in the **Identity** section, you need to select **Azure role assignments** and then **Add role assignments** to assign an Azure permission to the managed identity.

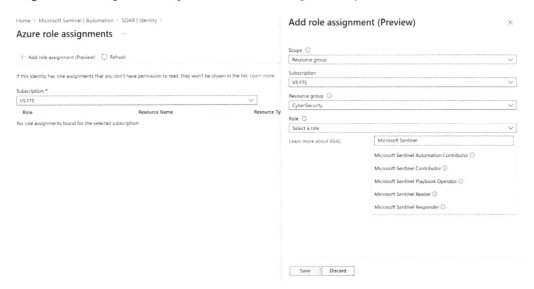

Figure 5.33 – Assigning a permission to a playbook's managed identity

To authenticate a playbook trigger or action with a managed identity, first, fill in **Connection name**, then for connecting, select **System-assigned managed identity** from the drop-down list, and click **Create**.

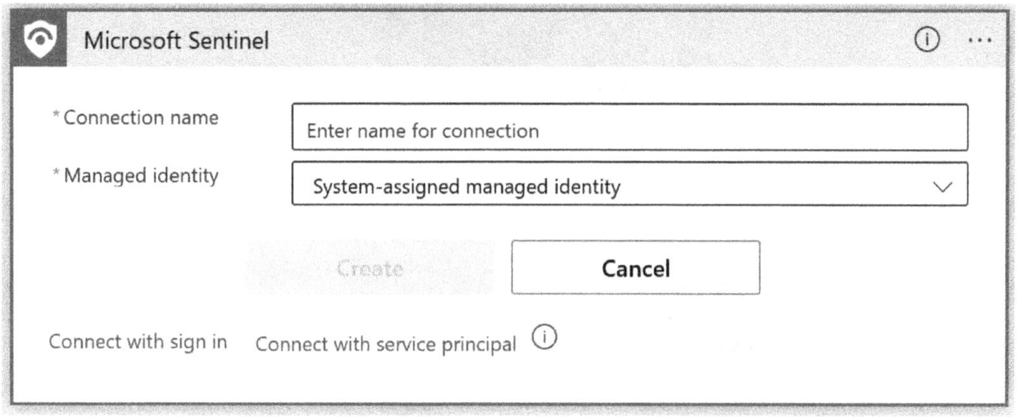

Figure 5.34 – Connecting to a Logic Apps connector using a managed identity

Service principals

Service principals can be created on the Azure AD administrator page by registering an application. Once we have created an application, we will need to save a **Tenant ID** and **Application ID** in a secure space. These will be needed for authentication purposes, as well as to create a secret for the application and save it in a secure space, such as Azure Key Vault.

To register an application, we need to enter the application's name and specify whether the application is single- or multi-tenant.

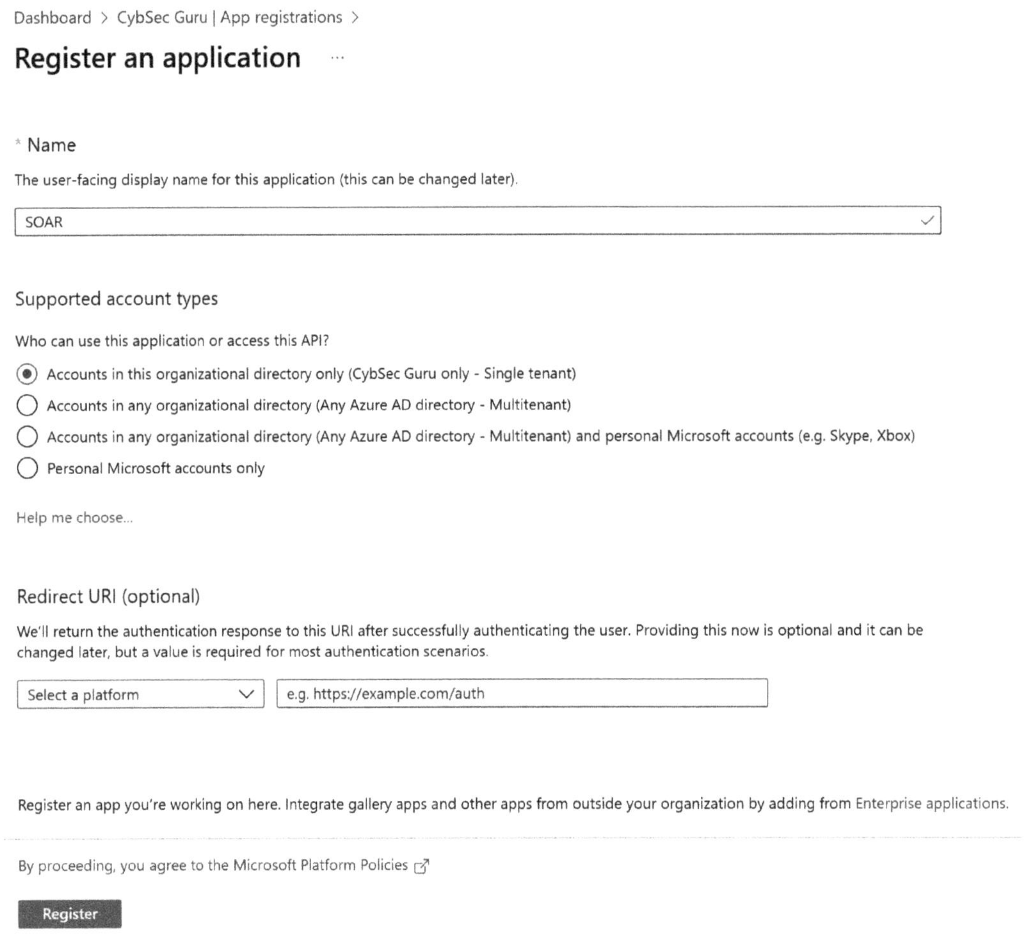

Dashboard > CybSec Guru | App registrations >

Register an application   ...

˟ Name

The user-facing display name for this application (this can be changed later).

| SOAR | ✓ |

Supported account types

Who can use this application or access this API?

- ◉ Accounts in this organizational directory only (CybSec Guru only - Single tenant)
- ○ Accounts in any organizational directory (Any Azure AD directory - Multitenant)
- ○ Accounts in any organizational directory (Any Azure AD directory - Multitenant) and personal Microsoft accounts (e.g. Skype, Xbox)
- ○ Personal Microsoft accounts only

Help me choose...

Redirect URI (optional)

We'll return the authentication response to this URI after successfully authenticating the user. Providing this now is optional and it can be changed later, but a value is required for most authentication scenarios.

| Select a platform ⌄ | e.g. https://example.com/auth |

Register an app you're working on here. Integrate gallery apps and other apps from outside your organization by adding from Enterprise applications.

By proceeding, you agree to the Microsoft Platform Policies ↗

Register

Figure 5.35 – Creating a new service principal using Azure AD App registrations

Once the service principal has been created, we can assign API permissions to it, such as the Microsoft Graph Security API and Microsoft 365 Defender.

Figure 5.36 – Assigning API permissions to a service principal

We can also assign any Azure AD or Azure **Role Based Access Control** (**RBAC**) role to a service principal. For example, suppose we want the option to query data in a Log Analytics workspace. In that case, we can utilize the **Azure Monitor Logs** connector, which does not support a managed identity at the time of writing but does support service principals. Therefore, we will assign the Log Analytics Reader permission to the service principal and authenticate our connector with it. Another important note about using a service principal is that it can be reused across multiple playbooks as it is not playbook-specific. This is because it's a system-assigned managed identity.

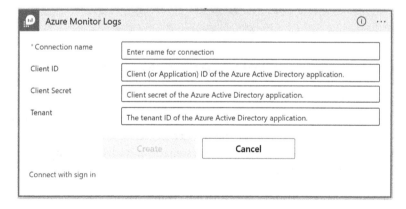

Figure 5.37 – Connecting to the Logic Apps connector using a service principal

User identity

The last option that we will cover is user identity. This option allows any user from your organization to authenticate the connection. The user carrying out the authentication must have the permission to perform the needed action (for example, to block a user in Azure Active Directory or to isolate a host in EDR) to authenticate the API connection. If the user doesn't have the permission, the playbook will fail on this step. This is also a shared connection. Once we create this API connection, it can be utilized across multiple playbooks – by a specific user or any user with permission to create and edit playbooks.

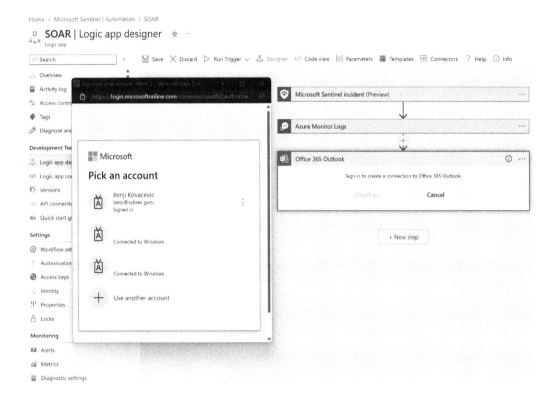

Figure 5.38 – Connecting to the Logic Apps connector using a user's identity

What's important to note here is that the user who authenticates this connection must have permission to perform the action (for example, to block a user in Azure Active Directory or isolate a host in EDR) when the playbook is running. Some organizations use **Privileged Identity Management** (**PIM**) and users are only assigned permission when needed. In this scenario, it is advisable to not use user identity for any action that can run while the user doesn't have permission active.

With that, let's shift focus and get to know triggers a little better.

Triggers

Triggers are used to define on what event a playbook will be triggered to run. It is always the first step when we create our playbook. Triggers also define the scheme that the playbook expects when it is triggered.

In Microsoft Sentinel, we have two primary triggers, with the third one still under development at the time of writing and, therefore, it will not be covered in this book.

The following triggers are available in Microsoft Sentinel:

- **Microsoft Sentinel alert**: This receives alert data as input

- **Microsoft Sentinel incident**: This receives incident data as input

- **Microsoft Sentinel entity**: Under development

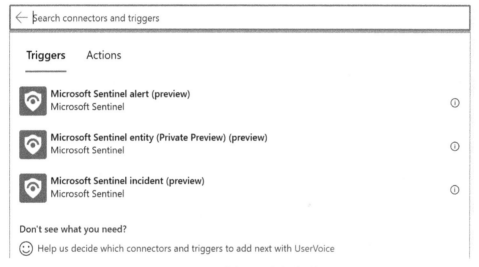

Figure 5.39 – Microsoft Sentinel playbook triggers

Since Microsoft Sentinel utilizes Logic Apps to create playbooks, we can utilize many more triggers that are not connected to Microsoft Sentinel. Some of them are as follows:

- **Schedule triggers**: Runs a playbook every *n* minutes, hours, or days, or at a specific time on specific days

- **Microsoft Forms triggers**: Creates a manual incident when a Microsoft Forms form is submitted

- **Office 365 Outlook**: Creates a manual incident when an email is received in a shared mailbox, and so on

Let's now focus on the rest of the playbook steps.

Actions

Once we have configured a trigger, we need to configure actions that will be performed when a playbook is triggered. We can utilize data that arrives with the trigger (incident data alongside the incident trigger, for example) to focus on specific information from it. These actions can be run sequentially, in parallel, or under complex conditions.

Microsoft Sentinel's native actions, among others, include the following:

- **Add comment to incident**
- **Bookmarks**: Create or update a bookmark
- **Create incident**
- **Entities**: Get accounts/hosts/IPs/URLs
- **Watchlist**: Create a new watchlist with data (raw content) and so on

Figure 5.40 – Microsoft Sentinel playbook actions

It's important to know that each action defines its scheme, and we can utilize data from triggers and previous actions to define data in the action we are working on now. To access this data, we can utilize dynamic content. This will be covered in the next section.

Some other actions we can create are as follows:

- **Azure Active Directory**: An action to block a user, reset their password, and revoke the session
- **Azure Firewall**: An action to block an IP
- **Palo Alto/Fortinet/CheckPoint…**: An action to block an IP
- **ServiceNow/Jira**: Sends incident data and creates a record
- **VirusTotal**: Gets the IP/URL/file hash enrichment data

Some other actions native to Logic Apps that we need to be aware of are as follows:

- **Condition**: Here, we can define which block of actions to execute based on condition evaluation. For example, if the host starts with *admin*, we can auto-close the incident; if not, we can isolate the device.
- **For each**: If the data that we are working on is an array (set of data) and we want to act on each piece of data in the dataset.
- **Switch**: Similar to **Condition**, but we can have multiple paths.
- **HTTP**: To make an API call to a product or service if a native action is not available in Logic Apps.
- **Parse JSON**: To parse a result received by an HTTP call.
- **Create HTML table**: To be used when creating an email response and you want to create an HTML table containing data.
- **Set variable**: If we want to set a variable that can be used on multiple instances in a playbook.

We will cover these actions in more depth in *Chapter 9*.

To make playbook actions utilize dynamic data, we can utilize dynamic content in Logic Apps.

Dynamic content

Dynamic content refers to temporary fields in a playbook run; these are created by triggers and actions. The only rule is that using these temporary fields is only possible for triggers and actions that happened before the action we are working on occurred. For example, when a playbook is triggered with a Microsoft Sentinel incident trigger, the output that's received will contain all the necessary data about the incident, such as its severity, status, incident number, incident URL, entities, and alerts. Using dynamic content, we can specify these values in the next action. For example, we might want to get a list of IPs, so we can use the **Entities - Get IPs** action; as input, we can use the **Entities** dynamic content that we received from the Microsoft Sentinel incident trigger.

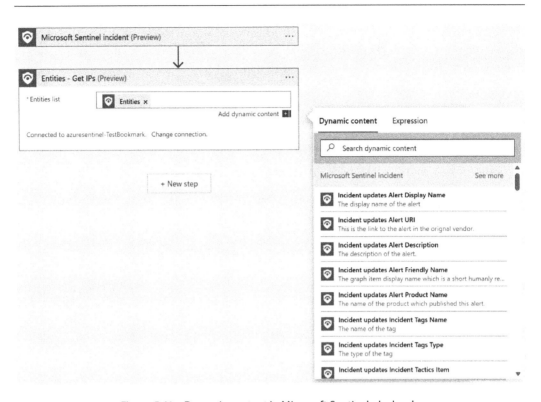

Figure 5.41 – Dynamic content in Microsoft Sentinel playbooks

If some data is not exposed in **Dynamic content** or we want to join two values, we can use expressions. An example of an expression is when we have an array containing alerts and we want to list alert names from that dataset; we can use a **join** expression to perform this action.

We will cover more on dynamic content and expressions in the hands-on examples in *Chapters 6* to *9*, which will feature common tips and tricks for working with Microsoft Sentinel playbooks.

Monitoring automation rules and playbook health

Microsoft Sentinel automation has a native way of monitoring the health of automation rules and playbook triggers. This monitoring can be enabled from the **Settings** page of Microsoft Sentinel, under **Health monitoring**.

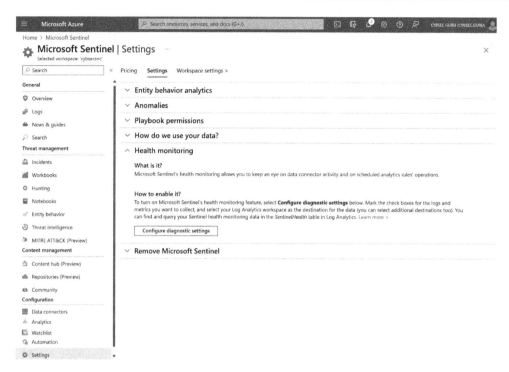

Figure 5.42 – Microsoft Sentinel – Health Monitoring configuration

We need to enable the **Automation** diagnostic settings and send them to the Log Analytics workspace where Microsoft Sentinel is enabled.

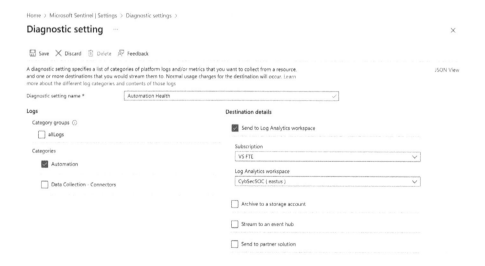

Figure 5.43 – The Automation Health monitoring configuration wizard

These diagnostic settings will be saved in the **SentinelHealth** table in Microsoft Sentinel so that we can query statuses using KQL.

A sample KQL query that you can use to get this data is as follows:

```
SentinelHealth
| where SentinelResourceType in ("Playbook", "Automation rule")
```

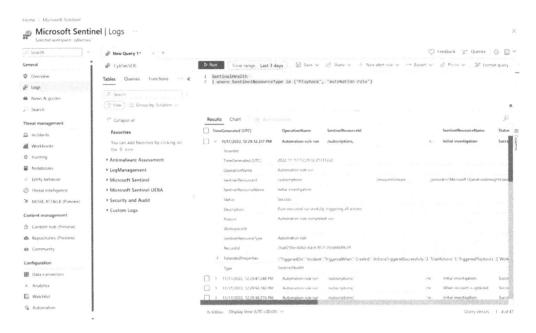

Figure 5.44 – Querying the SentinelHealth table in the Microsoft Sentinel | Logs tab

We can also utilize the **Automation Health** workbook that's available in workbook templates to get a detailed report about our automation health in Microsoft Sentinel.

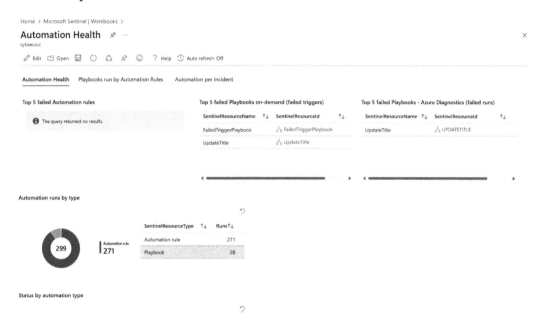

Figure 5.45 – Utilizing the Microsoft Sentinel Automation Health workbook

Automation Health only monitors a playbook trigger – in other words, it only checks whether the playbook triggered successfully, not the whole playbook run. To monitor whether the playbook runs successfully or not, we have to utilize diagnostic settings at the playbook level. A diagnostic setting can be configured when creating a playbook.

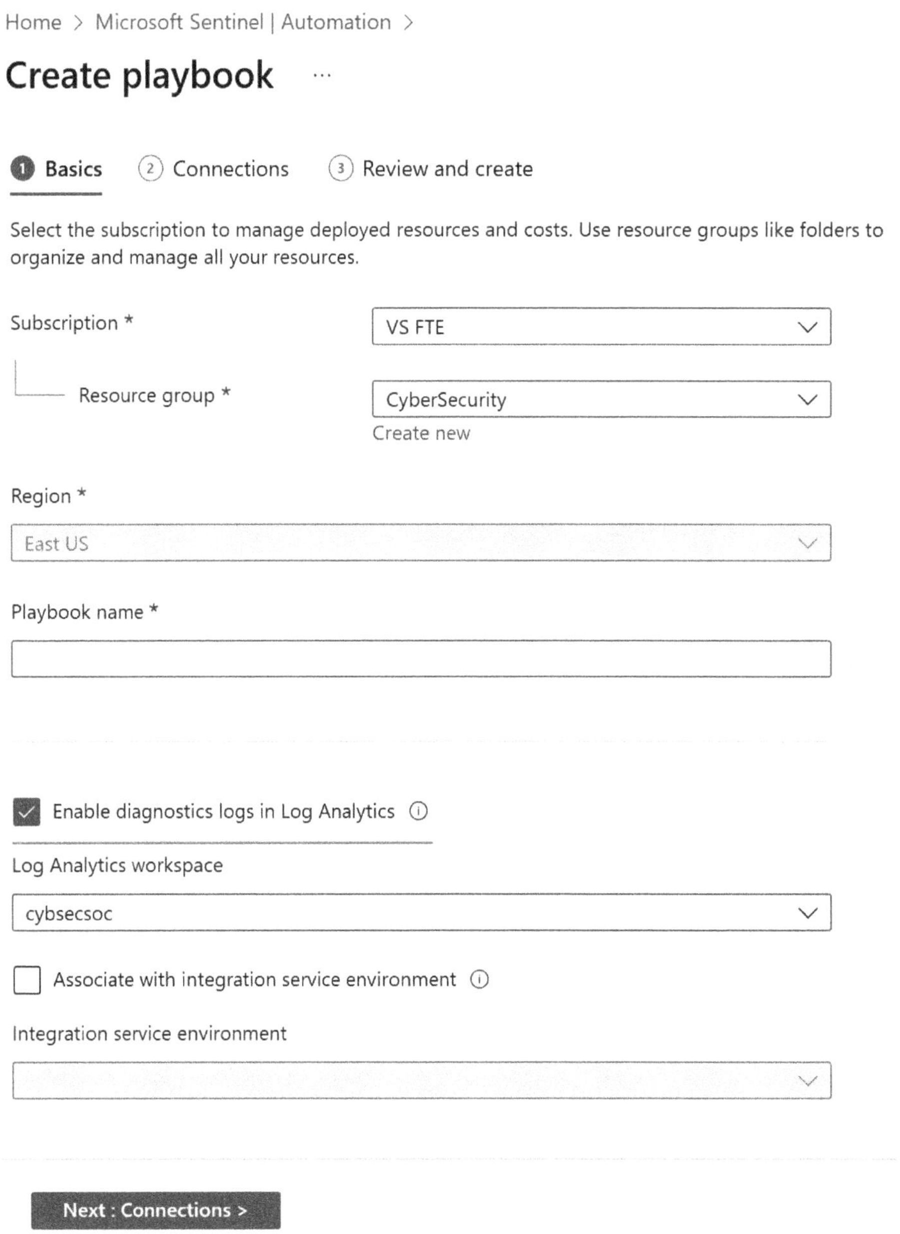

Home > Microsoft Sentinel | Automation >

Create playbook   ...

1 Basics   (2) Connections   (3) Review and create

Select the subscription to manage deployed resources and costs. Use resource groups like folders to organize and manage all your resources.

Subscription *

| VS FTE | ∨ |

Resource group *

| CyberSecurity | ∨ |

Create new

Region *

| East US | ∨ |

Playbook name *

| |

☑ Enable diagnostics logs in Log Analytics ⓘ

Log Analytics workspace

| cybsecsoc | ∨ |

☐ Associate with integration service environment ⓘ

Integration service environment

| | ∨ |

Next : Connections >

Figure 5.46 – Enabling diagnostic settings when creating a new playbook

A diagnostic setting can also be created in the playbook itself.

Figure 5.47 – Adding a diagnostic setting to an existing playbook

Once we've done this, we need to select the **Send to Log Analytics workspace** box, where we have Microsoft Sentinel enabled.

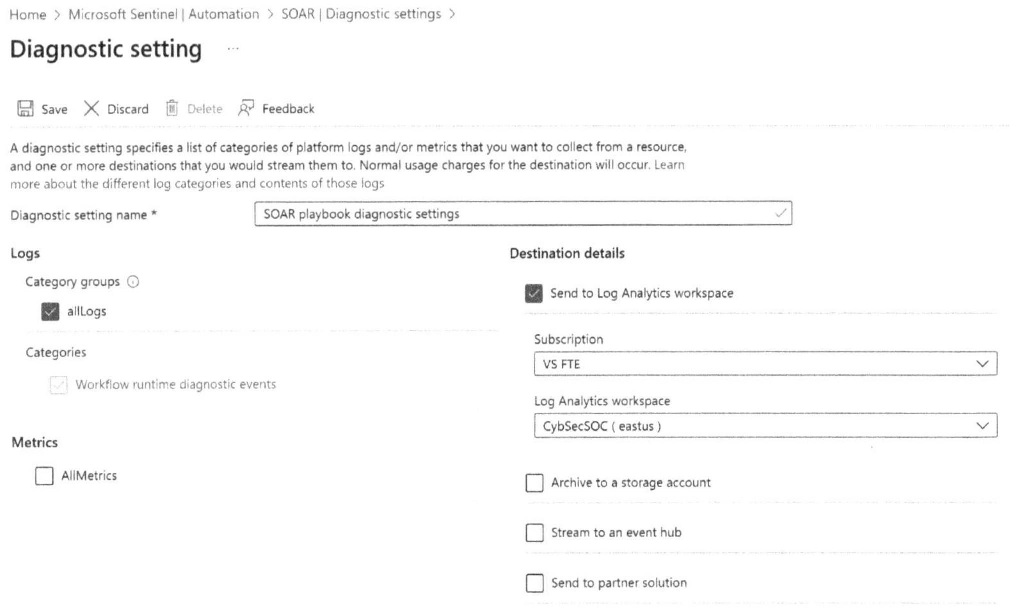

Home > Microsoft Sentinel | Automation > SOAR | Diagnostic settings >

Diagnostic setting   ...

🖫 Save   ✕ Discard   🗑 Delete   ⋊ Feedback

A diagnostic setting specifies a list of categories of platform logs and/or metrics that you want to collect from a resource, and one or more destinations that you would stream them to. Normal usage charges for the destination will occur. Learn more about the different log categories and contents of those logs

Diagnostic setting name *   [SOAR playbook diagnostic settings   ✓]

Logs   **Destination details**

  Category groups ⓘ   ☑ Send to Log Analytics workspace

    ☑ allLogs   Subscription
                                                                [VS FTE   ⌄]
  Categories
    ☑ Workflow runtime diagnostic events   Log Analytics workspace
                                                                [CybSecSOC (eastus)   ⌄]

Metrics

    ☐ AllMetrics   ☐ Archive to a storage account

                                                                ☐ Stream to an event hub

                                                                ☐ Send to partner solution

Figure 5.48 – The Diagnostic setting configuration wizard of a playbook

This data will be saved in the **AzureDiagnostics** table. We can query it using KQL:

```
AzureDiagnostics
| where OperationName == "Microsoft.Logic/workflows/
workflowRunCompleted"
```

The preceding KQL query will give you the following output:

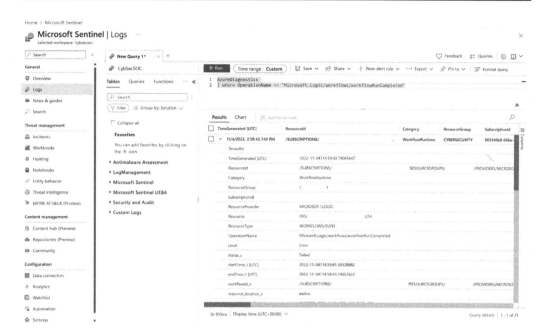

Figure 5.49 – The queried playbook's diagnostic settings in the Microsoft Sentinel | Logs tab

We can also join and compare data from the **AzureDiagnostics** and **SentinelHealth** tables to check whether the playbook triggered by the automation rule or the one we triggered manually had a successful run. This can be done by comparing the **runId** columns in both tables and joining them on the same **runId** since it is unique for each playbook run.

Here is a KQL example:

```
SentinelHealth
| where SentinelResourceType == "Automation rule"
| mv-expand TriggeredPlaybooks = ExtendedProperties.TriggeredPlaybooks
| extend runId = tostring(TriggeredPlaybooks.RunId)
| join (AzureDiagnostics
    | where OperationName == "Microsoft.Logic/workflows/
workflowRunCompleted"
    | project
        resource_runId_s,
        playbookName = resource_workflowName_s,
        playbookRunStatus = status_s)
    on $left.runId == $right.resource_runId_s
| project
    RecordId,
    TimeGenerated,
    AutomationRuleName= SentinelResourceName,
    AutomationRuleStatus = Status,
```

```
Description,
workflowRunId = runId,
playbookName,
playbookRunStatus
```

The preceding KQL query will give you the following output:

Figure 5.50 – Querying AzureDiagnostics and SentinelHealth in the Microsoft Sentinel | Logs tab

With automation rule and playbook health monitoring covered, let's wrap up this chapter.

Summary

In this chapter, we dug deep into Microsoft Sentinel automation and dissected each element. First, we focused on automation rules and their main elements – triggers, conditions, and actions – and how they define automation rule runs. We also covered permissions and ways to create automation rules. Then, we moved on to the topic of playbooks, where we focused on their main elements – triggers, actions, and dynamic content – as well as underlying information such as connectors, permissions, and authentication methods.

At the end of this chapter, we focused on the critical topic of automation health and how to monitor it using Microsoft Sentinel functionalities.

In the next chapter, we will begin our hands-on examples. We will focus on enriching incidents so that we can speed up MTTA and MTTR in Microsoft Sentinel.

6

Enriching Incidents Using Automation

In the previous chapter, we introduced Microsoft Sentinel automation and its main elements, permissions, and building blocks.

In this chapter, we will work through some hands-on examples. But first, we will guide you on how to enable Microsoft Sentinel to perform these exercises on your own, then we will go through our two hands-on examples – the enrichment of incidents with IP and URL details.

This chapter will go through the following topics:

- Why should you use automation for incident enrichment?
- Creating your own Microsoft Sentinel trail
- VirusTotal playbook – IP enrichment
- VirusTotal playbook – URL enrichment

Why should you use automation for incident enrichment?

When a new incident/case is detected, we first need to triage to check whether it is a true or false positive. In many cases, there are usually steps that SOC analysts perform when investigating new incidents. One example is if there is an IP address involved in the incident, SOC analysts will go to the **Threat Intelligence** blade to see whether it is malicious, or to external products such as **VirusTotal** or **Microsoft Defender Threat Intelligence** (**MDTI**), also known as **RiskIQ**. This takes SOC analysts' time, and they need to do it every time there is an IP involved in an incident, or more than one IP, which can take a few minutes for each incident. With the number of incidents increasing daily, this can mean hours lost in the SOC every day.

But what if we can perform that step even before SOC analysts pick up the incident for investigation?

This is done by creating a playbook that will run on incident/alert creation and enrich incidents with valuable information. That playbook will send requests to the enrichment system we want (such as MDTI or VirusTotal) and return feedback. In our case, it will be written in the comment section of the incident.

When SOC analysts pick up an incident for investigation, enrichment will already have been done, and the SOC analysts will not lose minutes visiting those portals themselves. This will save time in the SOC, but most importantly, it will improve MTTA and MTTR.

Creating your own Microsoft Sentinel trail

Before we start with hands-on examples, let's first go through the process to get access to **Microsoft Sentinel**. First, we need to create a Microsoft Azure account, as Microsoft Sentinel is a Microsoft Azure service:

1. To create a free Microsoft Azure account, please visit `https://azure.microsoft.com/en-in/free/`.

    When you enable Microsoft Azure, you get certain benefits for the first 12 months plus USD 200 credit for 30 days, which you can utilize to test certain Microsoft Azure services – in this case, Microsoft Sentinel.

2. Once you get access to Microsoft Azure, open the **Microsoft Azure** portal: `https://portal.azure.com/`.

3. In the search field, search for `Microsoft Sentinel`, and from the search results, choose **Microsoft Sentinel**, as shown in the following screenshot:

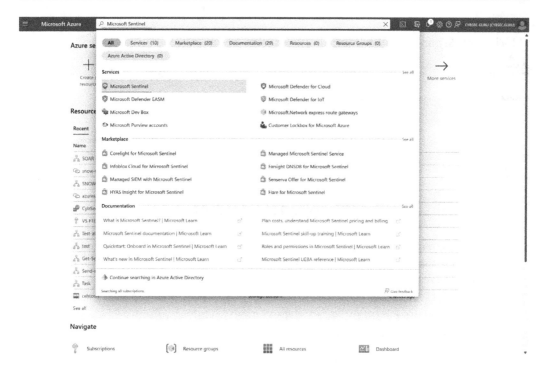

Figure 6.1 – Search for Microsoft Sentinel in the Microsoft Azure portal

4. In the next window, select **Create**:

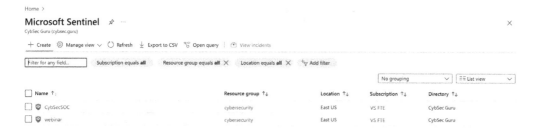

Figure 6.2 – Create a new Microsoft Sentinel environment

5. Since Microsoft Sentinel utilizes Log Analytics workspaces to store logs, we need to create a Log Analytics workspace to enable Microsoft Sentinel on top of it, as shown in the following screenshot:

Figure 6.3 – Create a new Log Analytics workspace

6. In the next window, create a new **Resource group**, enter the workspace's name, and select the region where the workspace will be deployed. This is the region where logs will be stored, as shown in *Figure 6.4*:

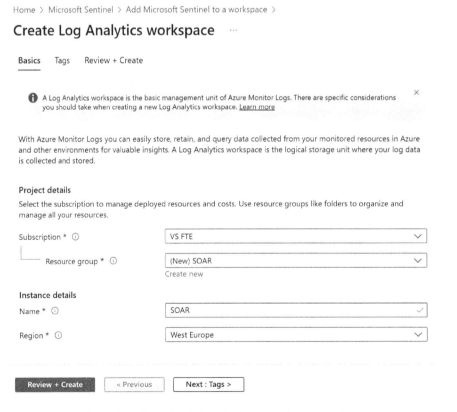

Figure 6.4 – Basics tab in Log Analytics workspace creation

7.   Under **Review + Create**, check the details and select **Create:**

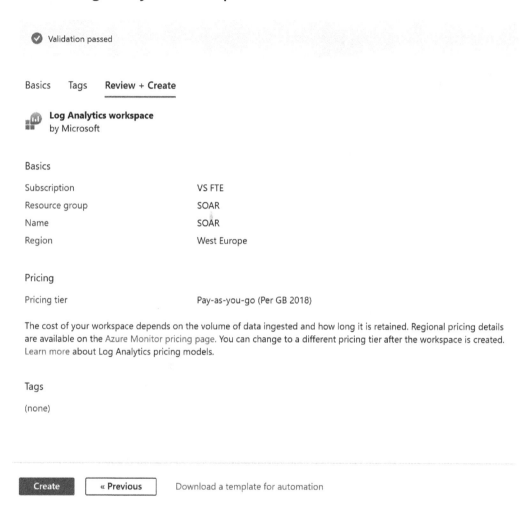

Home > Microsoft Sentinel > Add Microsoft Sentinel to a workspace >

Create Log Analytics workspace   ...

✓ Validation passed

Basics   Tags   **Review + Create**

Log Analytics workspace
by Microsoft

Basics

Subscription	VS FTE
Resource group	SOAR
Name	SOAR
Region	West Europe

Pricing

Pricing tier	Pay-as-you-go (Per GB 2018)

The cost of your workspace depends on the volume of data ingested and how long it is retained. Regional pricing details are available on the Azure Monitor pricing page. You can change to a different pricing tier after the workspace is created. Learn more about Log Analytics pricing models.

Tags

(none)

Create   « Previous   Download a template for automation

Figure 6.5 – Review and create a Log Analytics workspace

8. Once our Log Analytics workspace is deployed, we need to select it and click on **Add**:

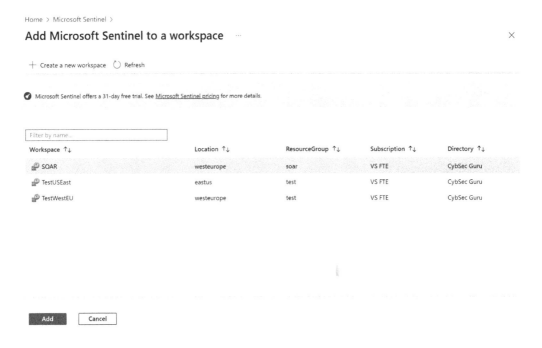

Figure 6.6 – Enable Microsoft Sentinel on top of the Log Analytics workspace

> **Note**
> Microsoft Sentinel has a 30-day free trial for certain services. All terms and conditions of the free trial can be found at `https://azure.microsoft.com/en-gb/pricing/details/microsoft-sentinel/`.

9. Once we click on **Add**, we will enable Microsoft Sentinel on top of our Log Analytics workspace, leading us directly to the **Microsoft Sentinel** page, as shown in the following screenshot:

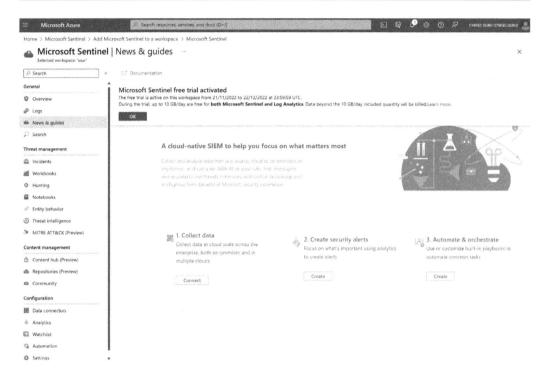

Figure 6.7 – Microsoft Sentinel News & guides tab

Now you are ready to start ingesting data and learn more about Microsoft Sentinel.

> **Note**
>
> You can visit the Microsoft Sentinel documentation and learn more about free sources you can ingest into Microsoft Sentinel here: `https://learn.microsoft.com/en-us/azure/sentinel/overview`.
>
> There is also step-by-step guidance that you can utilize to learn more about Microsoft Sentinel and how to navigate through the environment: `https://github.com/Azure/Azure-Sentinel/tree/master/Solutions/Training/Azure-Sentinel-Training-Lab`.

You can utilize **Microsoft Sentinel Training Lab** to populate your Microsoft Sentinel environment with demo data quickly. Go to **Content hub** from the left menu, search for `Training Lab`, and deploy it to your environment, as shown in *Figure 6.8*:

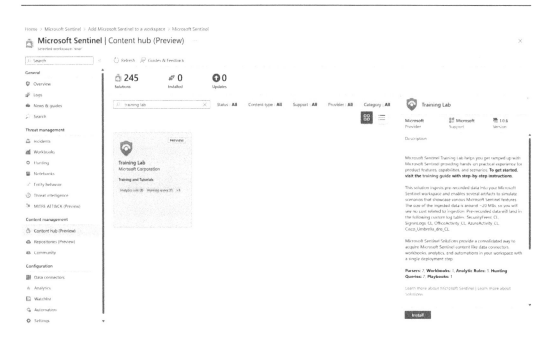

Figure 6.8 – Add Training Lab from the Content hub tab

Now that we have a Microsoft Sentinel environment we can test on, we can start creating our hands-on examples.

VirusTotal playbook – IP enrichment

In our first example, we will utilize VirusTotal to enrich the IP found in the incident.

Before we begin, you will need the following:

- You need to have access to Microsoft Sentinel with appropriate permissions (Microsoft Sentinel Contributor, Logic App Contributor, and permission to assign RBAC controls – Owner or User Access Administrator)

- You need to create a free VirusTotal account and get your API token for connector authentication or utilize a premium account if you have one

- If you don't have a VirusTotal account, you can create one at this link: `https://www.virustotal.com/gui/join-us`

- Once you have an account, you will need to get your API key, as shown in the following screenshot:

Figure 6.9 – Access to VirusTotal API key

For our demo, a standard API key is enough, as demonstrated in *Figure 6.10*. We will need this later to authenticate the API connection in the playbook.

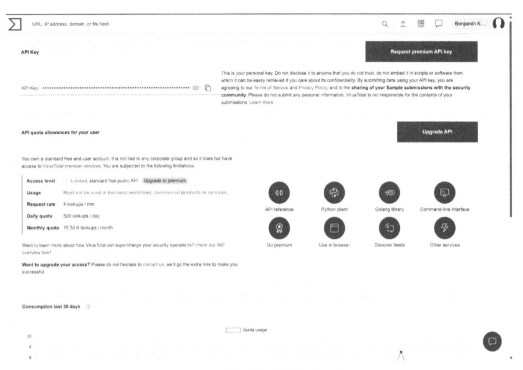

Figure 6.10 – VirusTotal API key location

If using in the production environment, consult VirusTotal about licensing terms and conditions.

Creating a playbook

Now that we have access to Microsoft Sentinel and a VirusTotal account and API key, we can start with our playbooks:

1. First, we need to open our Microsoft Sentinel environment, go to the **Automation** tab, click on **Create**, and select **Playbook with alert trigger** for this example, as shown in the following screenshot:

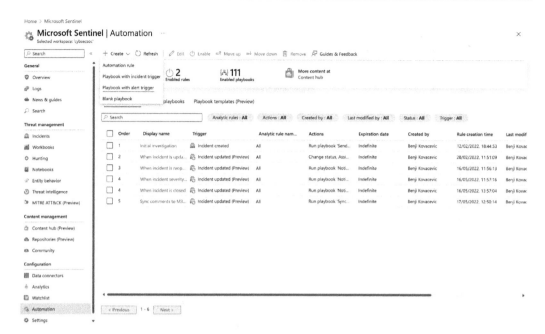

Figure 6.11 – Create a playbook with an alert trigger

2. In the **Basics** tab of playbook creation, select your resource group (existing or new) and enter your playbook name (in my case, `VirusTotal-IPEnrichment-alertTrigger`), as shown in the following screenshot:

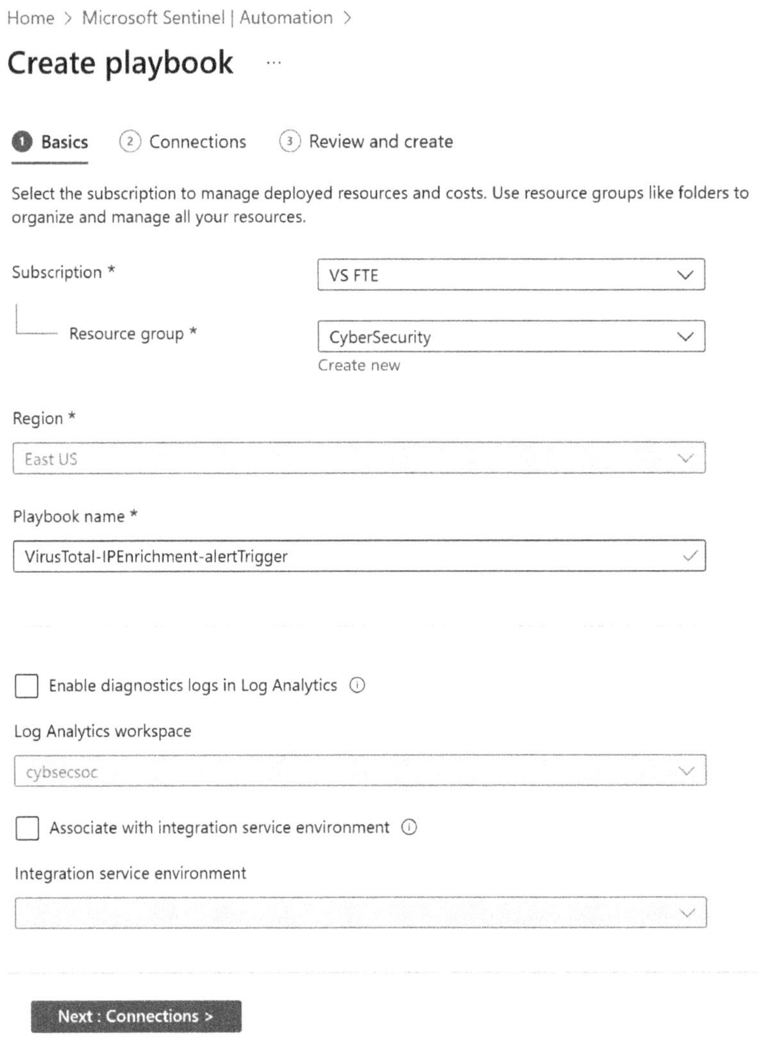

Home > Microsoft Sentinel | Automation >

Create playbook   ...

1 Basics   (2) Connections   (3) Review and create

Select the subscription to manage deployed resources and costs. Use resource groups like folders to organize and manage all your resources.

Subscription *

| VS FTE ⌄ |

└──── Resource group *

| CyberSecurity ⌄ |
Create new

Region *

| East US ⌄ |

Playbook name *

| VirusTotal-IPEnrichment-alertTrigger ✓ |

☐ Enable diagnostics logs in Log Analytics ⓘ

Log Analytics workspace

| cybsecsoc ⌄ |

☐ Associate with integration service environment ⓘ

Integration service environment

| ⌄ |

Next : Connections >

Figure 6.12 – Basics tab in the playbook creation wizard

3.   In the next window, **Connections**, we will leave it as it is, as shown in *Figure 6.13*:

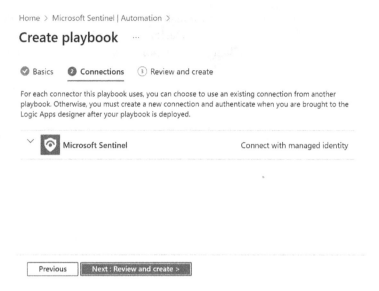

Figure 6.13 – Choose your connection for the Microsoft Sentinel connector

4. In **Review and create**, check the configuration and select **Create and continue to designer**, as shown in *Figure 6.14*:

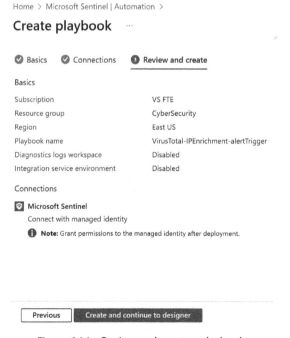

Figure 6.14 – Review and create a playbook

5. When our playbook is deployed, we will go through the process of changing the API connection for the Microsoft Sentinel connection. In the **Designer** view, click on the Microsoft Sentinel alert trigger to expand it and select **Change connection**, as shown in the following screenshot:

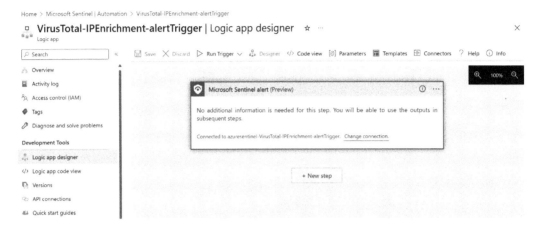

Figure 6.15 – Change connection in Logic app designer

6. We can select an existing connection from the list or select **Add new** to create a new one. In this case, we will select **Add new**, as shown in the following screenshot:

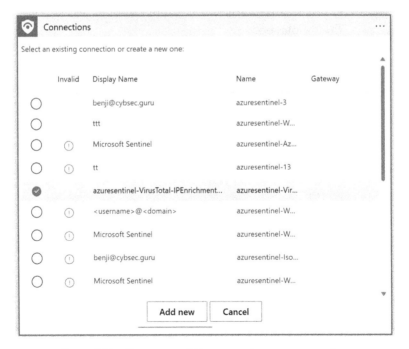

Figure 6.16 – Choose an existing or create a new connection

In the next window, we select between the different authentication options that we covered in *Chapter 5* – managed identity, service principal, or user identity, as shown in *Figure 6.17*:

Figure 6.17 – Available authentication options to create a new connection

7. We will use the user identity, so we will click on **Sign in**. After that, the Microsoft sign-in page will appear, and we'll need to sign in with our credentials, as shown in *Figure 6.18*:

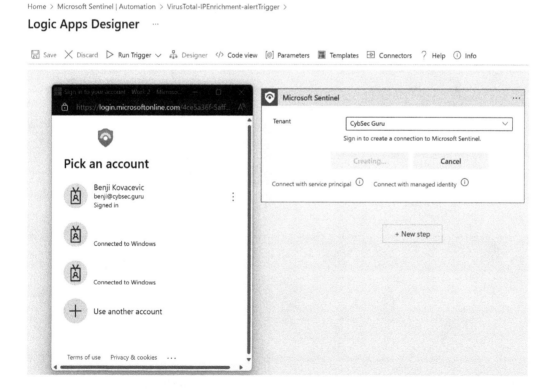

Figure 6.18 – The sign-in window when utilizing user identity

8.   Once we sign in, we will see a change in our connection, as shown in the following screenshot:

Figure 6.19 – User identity connection

9.   The next step in our playbook creation is to add a new action to get IPs from the incident. In our playbook design view, we will click on **New step**, as shown in the following screenshot:

Figure 6.20 – Add a new step

10. We will search for `Microsoft Sentinel` and select it from the results, as shown in the following screenshot:

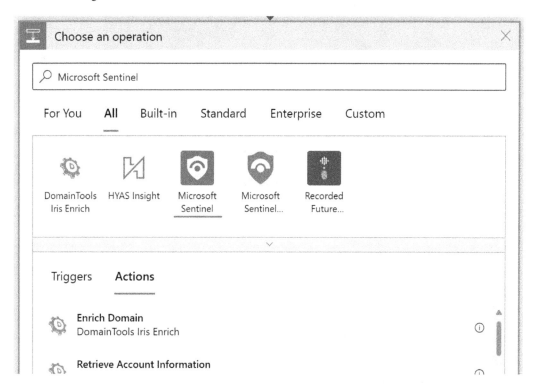

Figure 6.21 – Search for and select the connection

11. From our list of actions, search for and select **Entities - Get IPs**:

Figure 6.22 – Search for and select the action

12. The new action will be added, and we now need to specify in this action from where to extract those incidents. As we mentioned in *Chapter 5*, each trigger and action has its own scheme, and we can utilize dynamic content to read this data and always get the data needed for the actual playbook run, as shown in *Figure 6.23*:

Figure 6.23 – The Entities - Get IPs action

13. In this case, we will need to get IPs from our alert trigger. Click on an empty space in our **Entities - Get IPs** action, and you will see **Dynamic content** showing on the right. In this case, we will see only content from the Microsoft Sentinel alert trigger, as that is the only trigger/action that happened before the action we are working on now.

14. In the **Dynamic content** search field, we will search for `entities` and select them. This will add the **Entities** dynamic content to our **Entities - Get IPs** action, as shown in the following screenshot:

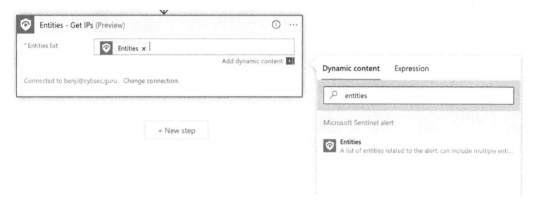

Figure 6.24 – Add Entities as dynamic content

15. Now that we have extracted IPs from other entities in our playbook, it's time to add a VirusTotal action. Click on **New step**, search for and select **Virus Total**, and select the **Get an IP report** action from the list, as shown in *Figure 6.25*:

Figure 6.25 – Add the Get an IP report action

16. We will get a window to name our connection (I will use `VirusTotal`) and to enter our API key. Once you've entered both, click on **Create**, as shown in *Figure 6.26*:

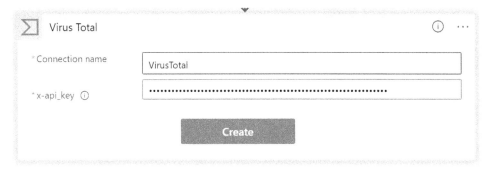

Figure 6.26 – Authenticate the VirusTotal connection with the API key

We will see our **Get an IP report** action from VirusTotal. In it, we need to specify the IP address for which we want to get the report. We will again use **Dynamic content** and get IP addresses from our **Entities - Get IPs** action. Click on text space, and from **Dynamic content**, search for **IPs Address** under **Entities - Get IPs**.

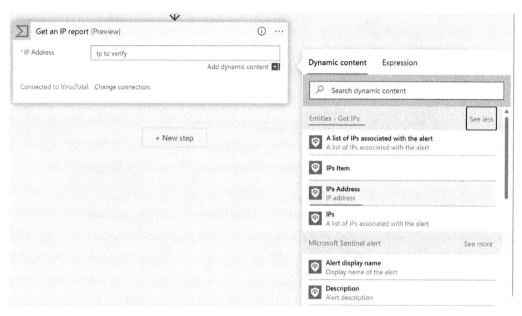

Figure 6.27 – Add IP address from Dynamic content

One thing that will happen is that the playbook will automatically create a **For each** condition, which will be based on IPs, as shown in the following screenshot:

Figure 6.28 – For each loop

This happens because we can have more than one IP address in our incident or alert (based on the trigger), and we need to check IP addresses one by one to see their reputation.

So we will send a request to VirusTotal for an IP report, which will be returned to us in response, but how can we use it in Microsoft Sentinel? We can add it as a comment or a tag. In this case, we will add it as a comment.

17. Instead of selecting **New step**, we need to select **Add an action** from the **For each** loop, as we need to add a comment for each IP address, as shown in the following screenshot:

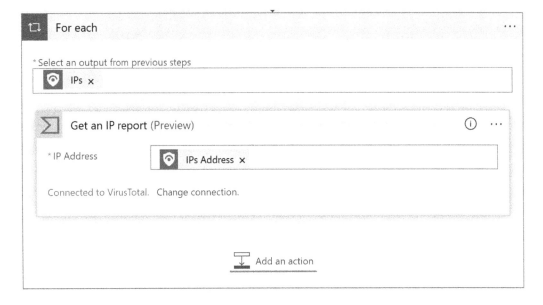

Figure 6.29 – Add an action in the For each loop

18. We will now see that we can utilize recently used connectors immediately, so we don't need to search for them but select them, as shown in *Figure 6.30*:

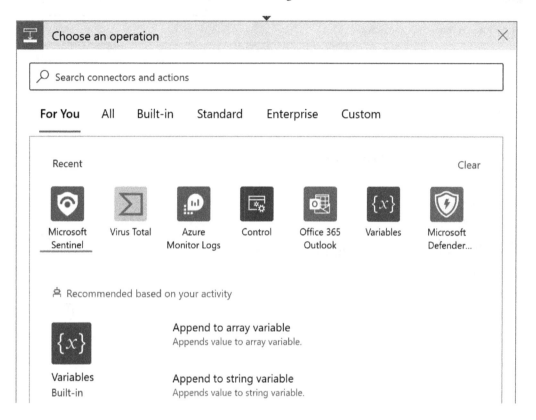

Figure 6.30 – Recent connections

19. From the list of actions, we will select **Add comment to incident**, as shown in the following screenshot:

Figure 6.31 – Add comment to incident action

20. We can now see that we are asked for an incident ARM ID, as Microsoft Sentinel doesn't have the option to leave a comment on the alert itself but on the incident. If we search for it in **Dynamic content**, we will not be able to find it, as shown in *Figure 6.32*:

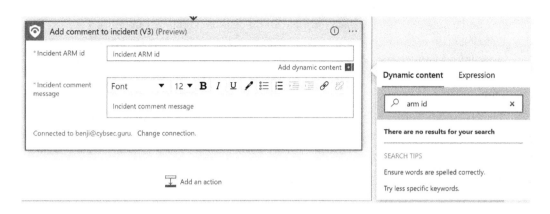

Figure 6.32 – No incident ARM ID in the alert trigger

So how will we get it?

One useful action when working with Microsoft Sentinel alert triggers in this scenario is called **Alert - Get incident**. This action will allow us to get information about the incident it is part of based on the alert we are investigating.

But where should we add it?

21. As we need to utilize **Dynamic content**, we must add it before the **Add comment to incident** action. We can add it inside the **For each** loop or even before the **For each** loop.

What is the difference?

If we add it inside the **For each** loop, that action will be run for each IP. We can do it by selecting the plus (+) sign between the **Get an IP report** and **Add comment to incident** actions. This sign will appear if we hover over the arrow between these actions, as shown in *Figure 6.33*:

Figure 6.33 – Insert a new step between two actions in the For each loop

If we add it outside of the **For each** loop, it will run once, and then we can utilize it across the playbook without rerunning it. This will impact pricing and speed, as the action will be run fewer times. In our case, it will not have any impact as we will test it on one IP address, and the cost of running an action in a Microsoft Sentinel playbook is minimal if you are not running it thousands of times per day.

22. To add this step outside the **For each** loop is the same process as within the **For each** loop, but we do it between actions outside the **For each** loop, as shown in the following screenshot:

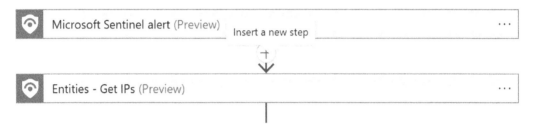

Figure 6.34 – Insert a new step between two actions before the For each loop

As it is best practice to use it as few times as possible, unless something different is required, we will use a process outside the **For each** loop.

23. Hover over the arrow between the actions, click on **Insert a new step**, and then on **Add an action**. From our recent connections, select **Microsoft Sentinel** and search for an action called **Alert - Get incident**, as shown in the following screenshot:

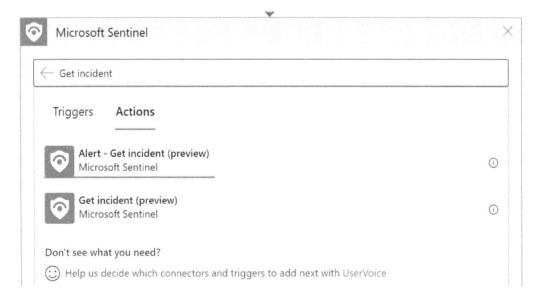

Figure 6.35 – Alert - Get incident action

24. When we add the action, we will need to add details about **Subscription ID**, **Resource group**, and **Workspace ID** where the incident is, as well as the alert ID. All this information is available as **Dynamic content** in our trigger, and we can select it easily, as shown in the following screenshot:

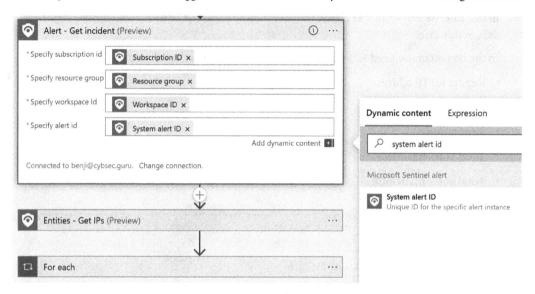

Figure 6.36 – Populate action data using Dynamic content

25. So, now we have our action to get incident details, and we can go back to the **For each** loop and **Add comment to incident**. If we again search for ARM ID in **Dynamic content**, we will see it on our result page, as shown in the following screenshot:

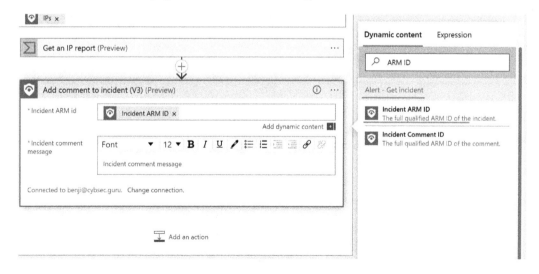

Figure 6.37 – Incident ARM ID is now available

26. The next step is to add the comment text itself. The great thing is that Microsoft Sentinel comments support HTML, so if needed, we can write HTML code in a friendlier format in Microsoft Sentinel.

In this case, we will write it as classic text. We want to know the IP owner and the last statistical data in this case.

In the comment, we need to write the following, as shown in the following screenshot:

- **Report for IP address**
- **Owner**
- **Reputation**
- **Total votes harmless**
- **Total votes malicious**

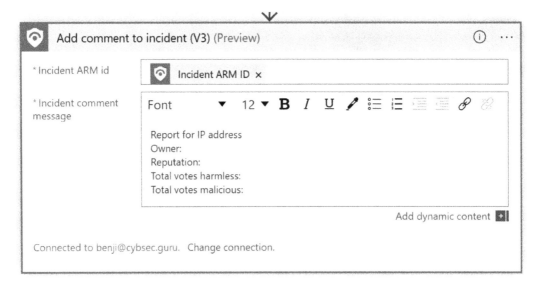

Figure 6.38 – Incident comment message text values

27. And now, we need to add dynamic data that will enrich our incident, as shown in *Figure 6.39*:

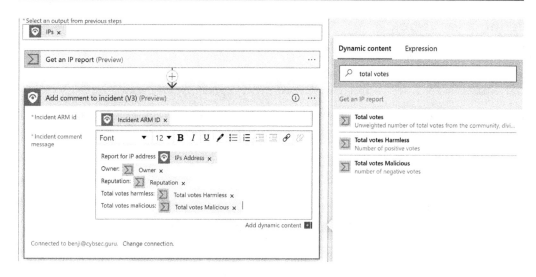

Figure 6.39 – Adding dynamic content from previous actions to the comment

28.   Now we need to select **Save** to save our playbook, as shown in *Figure 6.40*:

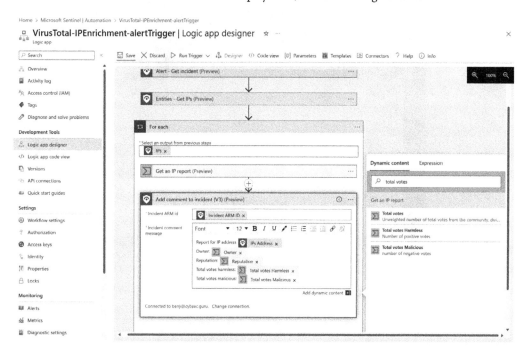

Figure 6.40 – Final playbook look and the Save option

And with that, our first playbook has been created!

Testing a playbook

Microsoft Sentinel will need an actual alert/incident to test the playbook. We can create a test Scheduled analytic rule to test it using the Microsoft Sentinel feature called **watchlists**. Watchlists allow us to create dynamic tables inside Microsoft Sentinel, where we can add/remove data.

To test our playbook, we will create a watchlist and ingest data from the **Comma-Separated Values (CSV)** file:

1. First, let's open a blank Microsoft Excel workbook and, in cell A1, insert IP, and in cell A2, insert 45.81.226.17, as shown in the following screenshot:

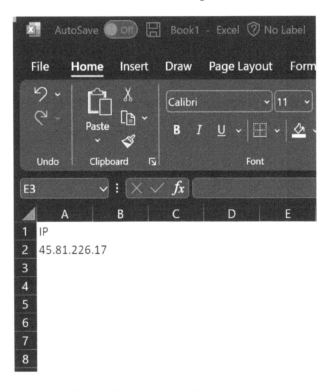

Figure 6.41 – Add data to the Excel table

2. Click on **File**, then **Save as**, choose a location, enter the name, and choose to save it as a CSV file, as shown in the following screenshot:

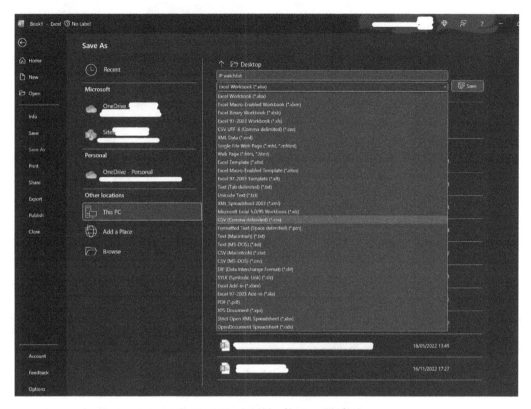

Figure 6.42 – Save the file as a CSV file

3. Now, go to **Microsoft Sentinel**, select the **Watchlist** tab, and click **Add new**, as shown in the following screenshot:

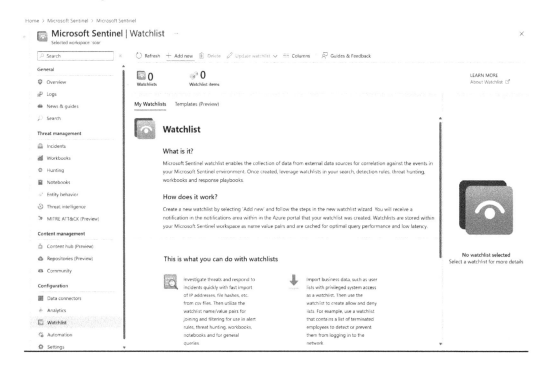

Figure 6.43 – Add a new watchlist from the Watchlist tab

4. In the **General** tab, enter the name and alias of the watchlist (`MaliciousIP`), and click on **Next: Source**, as shown in *Figure 6.44*:

Home > Microsoft Sentinel > Microsoft Sentinel | Watchlist >

Watchlist wizard   ···

Create new watchlist

General   Source   Review and create

Name *

| MaliciousIP | ✓ |

Description

| |

Alias *

| MaliciousIP | ✓ |

Next: Source >

Figure 6.44 – General tab in the watchlist wizard

5. Click on **Browse for files** in the next window and upload the saved CSV file. For the **SearchKey** value, select **IP** and click on **Next: Review and create**, as shown in the following screenshot:

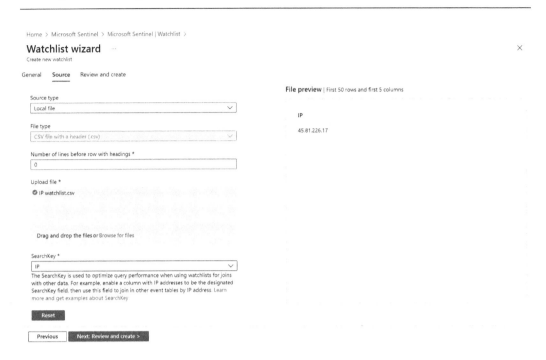

Figure 6.45 – Upload the file and select a SearchKey value

6.   Check your configuration and click on **Create**, as shown in the following screenshot:

Figure 6.46 – Review the configuration and create the watchlist

It will be around 5-10 minutes until this data is available for us to query.

7. After approximately 5 minutes, we can click **Refresh** to see whether the number of rows changed from 0 to 1. If it has, click on **View in logs**. If not, wait for a couple more minutes, as shown in the following screenshot:

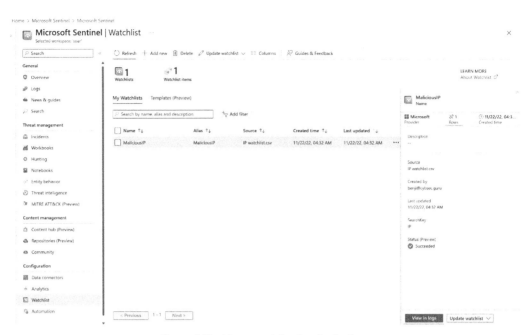

Figure 6.47 – View watchlist data in the logs

8. This will open this watchlist in **Logs**, and we can now create a Microsoft Sentinel Schedule rule from it by selecting **New alert rule** and **Create Azure Sentinel alert**, as shown in *Figure 6.48*:

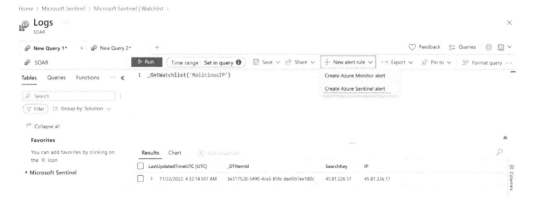

Figure 6.48 – Create a new detection rule from Logs

9. It will open the analytics rule creation wizard. (I named it `Test - Malicious IP`, and I left the rest unchanged.) Click on **Next: Set rule logic**, as shown in the following screenshot:

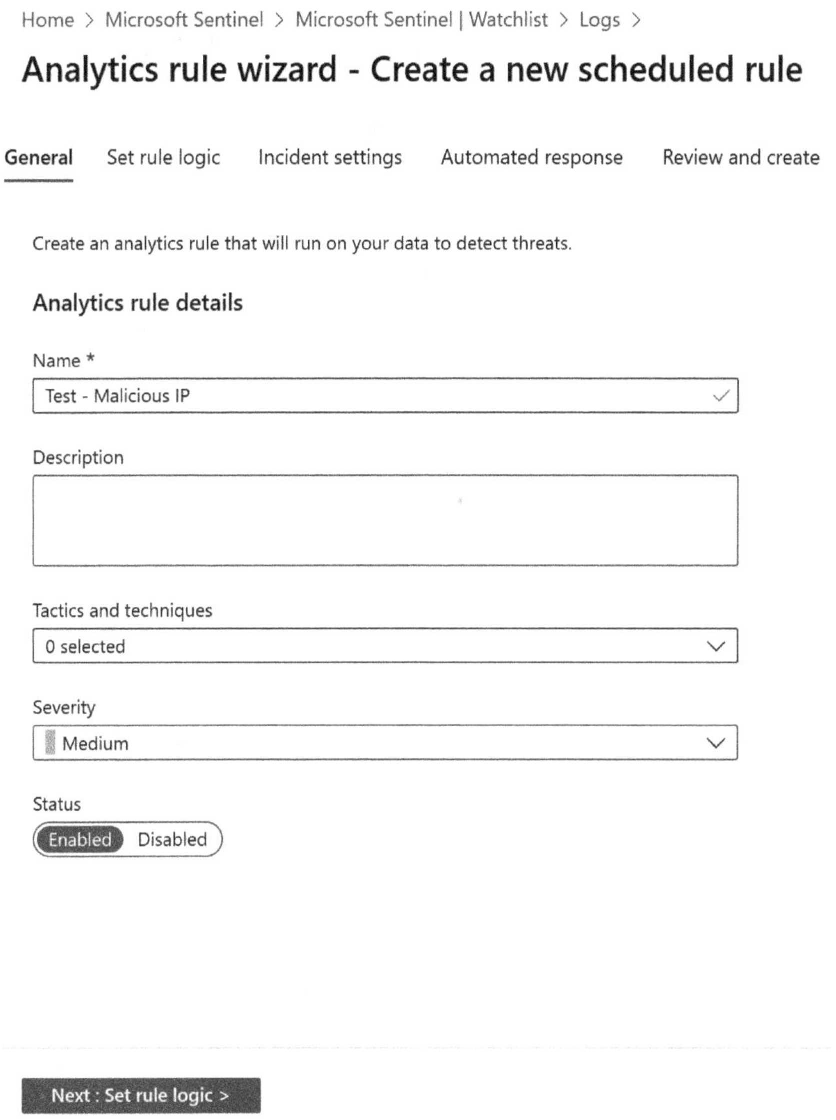

Home > Microsoft Sentinel > Microsoft Sentinel | Watchlist > Logs >

Analytics rule wizard - Create a new scheduled rule

General   Set rule logic   Incident settings   Automated response   Review and create

Create an analytics rule that will run on your data to detect threats.

Analytics rule details

Name *

Test - Malicious IP ✓

Description

Tactics and techniques

0 selected ⌄

Severity

Medium ⌄

Status

(**Enabled** Disabled)

Next : Set rule logic >

Figure 6.49 – Analytics rule wizard – General tab

10. In **Set rule logic**, we can see our query under **Rule query**. Now, we will set **Entity mapping** by selecting **IP** for the entity type, **Address** as the identifier, and **IP** as the value, as shown in the following screenshot:

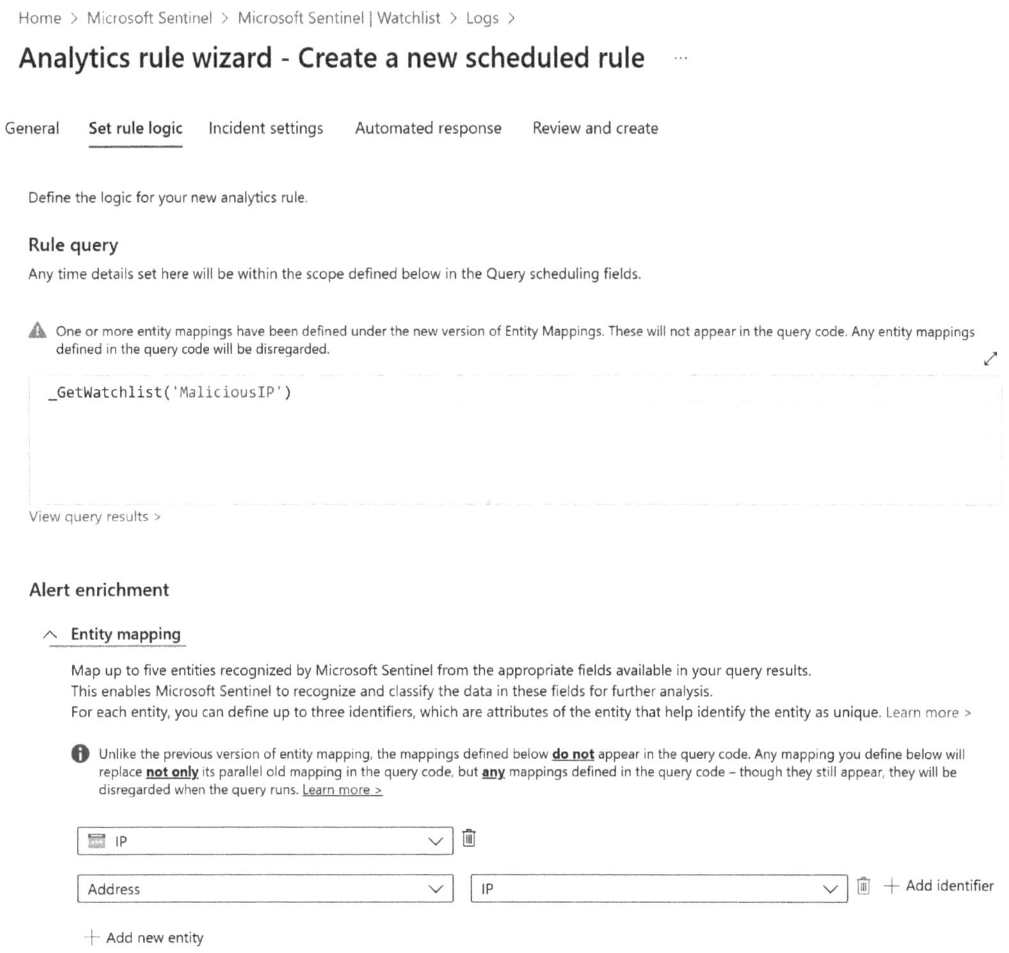

Home > Microsoft Sentinel > Microsoft Sentinel | Watchlist > Logs >

Analytics rule wizard - Create a new scheduled rule ⋯

General   **Set rule logic**   Incident settings   Automated response   Review and create

Define the logic for your new analytics rule.

Rule query

Any time details set here will be within the scope defined below in the Query scheduling fields.

⚠ One or more entity mappings have been defined under the new version of Entity Mappings. These will not appear in the query code. Any entity mappings defined in the query code will be disregarded.

```
_GetWatchlist('MaliciousIP')
```

View query results >

Alert enrichment

∧ **Entity mapping**

Map up to five entities recognized by Microsoft Sentinel from the appropriate fields available in your query results.
This enables Microsoft Sentinel to recognize and classify the data in these fields for further analysis.
For each entity, you can define up to three identifiers, which are attributes of the entity that help identify the entity as unique. Learn more >

ℹ Unlike the previous version of entity mapping, the mappings defined below **do not** appear in the query code. Any mapping you define below will replace **not only** its parallel old mapping in the query code, but **any** mappings defined in the query code – though they still appear, they will be disregarded when the query runs. Learn more >

| 🗒 IP | ∨ | 🗑 |

| Address | ∨ | IP | ∨ | 🗑 + Add identifier |

+ Add new entity

Figure 6.50 – Entity mapping in the Set rule logic tab

11. The last step in **Set rule logic** will be to change **Run query every** from **Hour** to **Minutes**, and then we can click on **Next: Incident settings**, as shown in the following screenshot:

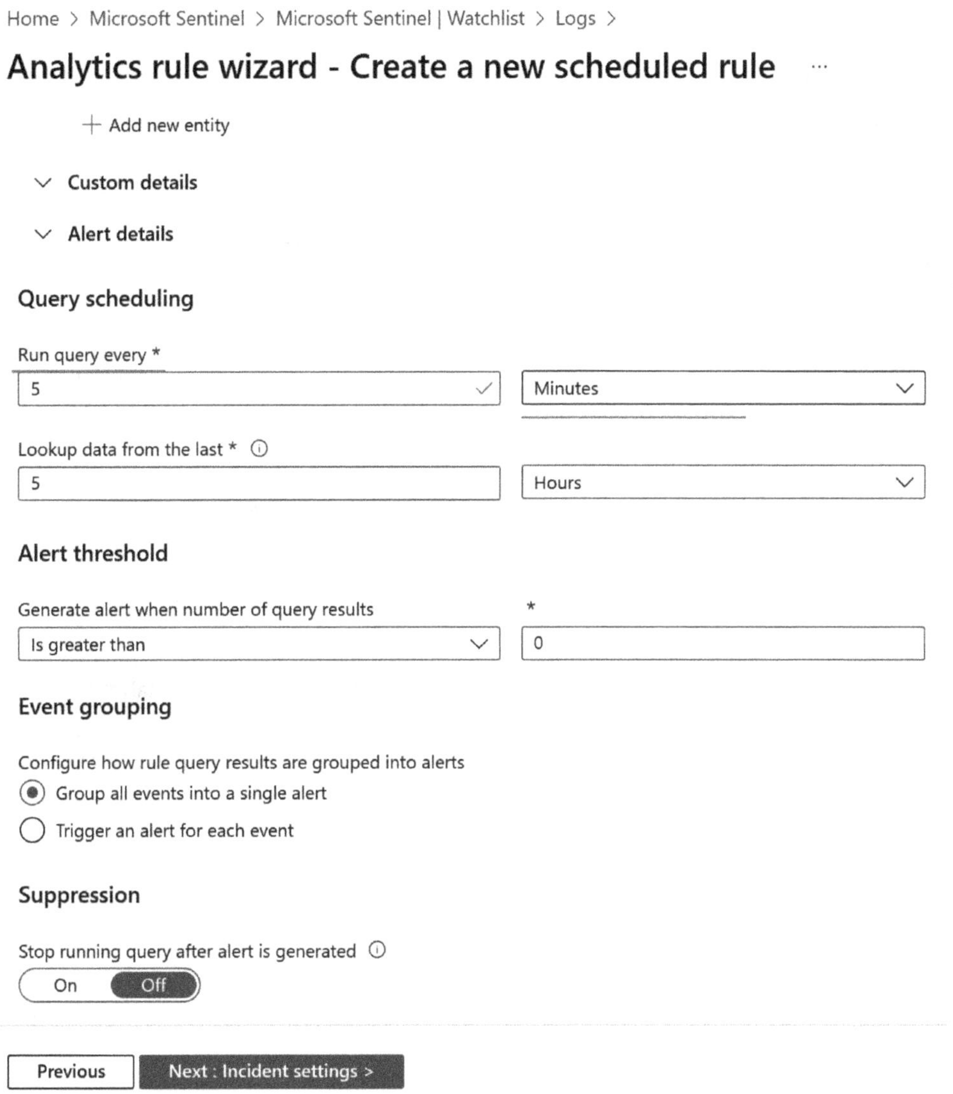

Home > Microsoft Sentinel > Microsoft Sentinel | Watchlist > Logs >

Analytics rule wizard - Create a new scheduled rule ···

+ Add new entity

∨ **Custom details**

∨ **Alert details**

Query scheduling

Run query every *

| 5 | ✓ | | Minutes | ∨ |

Lookup data from the last * ⓘ

| 5 | | | Hours | ∨ |

Alert threshold

Generate alert when number of query results   *

| Is greater than | ∨ | | 0 |

Event grouping

Configure how rule query results are grouped into alerts

⦿ Group all events into a single alert

◯ Trigger an alert for each event

Suppression

Stop running query after alert is generated ⓘ

| On | **Off** |

| Previous | | **Next : Incident settings >** |

Figure 6.51 – Run query scheduling configuration in the Set rule logic tab

12. We will leave the incident settings unchanged and continue to **Automated response**.

In **Automated response**, we will create an automation rule that will run when this analytics rule is created. We will click on **Add new**, enter the name, and select **When alert is created** as the trigger, as shown in *Figure 6.52*:

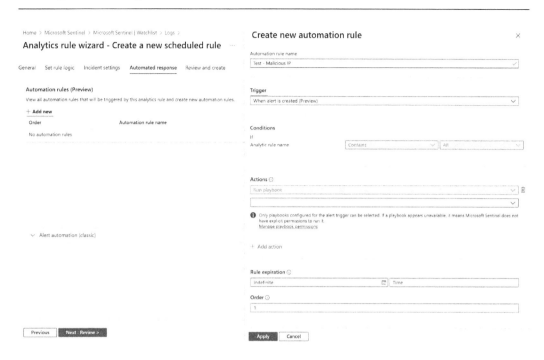

Figure 6.52 – Create a new automation rule from the Automated response tab

13. As an action, the only possibility is to run a playbook, but we will need to assign permission to Microsoft Sentinel to run a playbook, a role named Microsoft Sentinel Automation Contributor, which we covered in *Chapter 5*.

To assign it, we will need to click on **Manage playbook permissions**:

Figure 6.53 – Manage playbook permissions if not configured already

14. We will select our resource group where we saved the playbook and click on **Apply**:

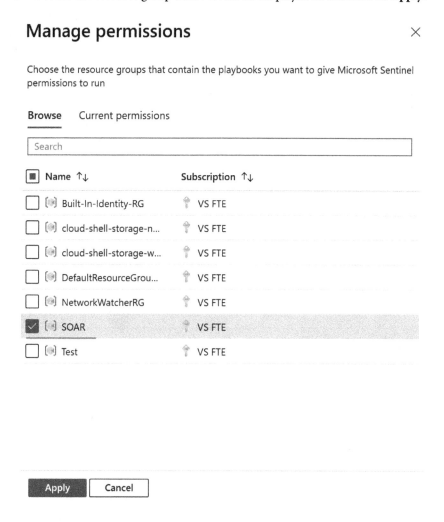

Manage permissions   ✕

Choose the resource groups that contain the playbooks you want to give Microsoft Sentinel permissions to run

Browse   Current permissions

Search

■ Name ↑↓	Subscription ↑↓
☐ Built-In-Identity-RG	🔑 VS FTE
☐ cloud-shell-storage-n...	🔑 VS FTE
☐ cloud-shell-storage-w...	🔑 VS FTE
☐ DefaultResourceGrou...	🔑 VS FTE
☐ NetworkWatcherRG	🔑 VS FTE
☑ SOAR	🔑 VS FTE
☐ Test	🔑 VS FTE

[Apply] [Cancel]

Figure 6.54 – Select one or more resource groups

15. We will now add the VirusTotal playbook we created, click on **Apply**, and we will see our automation rule with order number 1, as shown in the following screenshot:

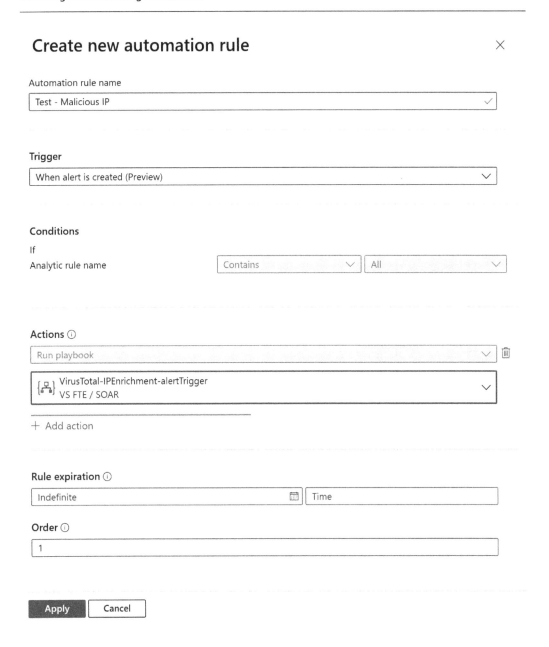

Figure 6.55 – Attach a playbook

16. Select **Next: Review** to continue to the last step in analytic rule creation, as shown in the following screenshot:

Analytics rule wizard - Create a new scheduled rule   ···

General   Set rule logic   Incident settings   **Automated response**   Review and create

Automation rules (Preview)

View all automation rules that will be triggered by this analytics rule and create new automation rules.

+ **Add new**

Order	Automation rule name	Trigger	Action	Status
1	Test - Malicious IP	🛡 Alert created (Preview)	Run playbook 'VirusTotal-IPEnrichment-...	⏻ Enabled ···

Figure 6.56 – The automation rule will appear in the view

17. In **Review and create**, we can check our configuration and click on **Create**.

Analytics rule wizard - Create a new scheduled rule   ···

✓ Validation passed.

General   Set rule logic   Incident settings   Automated response   **Review and create**

Analytics rule details

Name	Test - Malicious IP
Description	
Tactics and techniques	
Severity	▌ Medium
Status	⏻ Enabled

Analytics rule settings

Rule query	_GetWatchlist('MaliciousIP')
Rule frequency	Run query every **5 minutes**
Rule period	Last **5 hours** data
Rule start time	Automatic
Rule threshold	Trigger alert if query returns **more than 0** results
Event grouping	Group all events into a single alert
Suppression	Not configured

Entity mapping

Entity 1:	**IP**
	Identifier: Address, Value: IP

Previous   Create

Figure 6.57 – Review the configuration and create a new scheduled rule

This will create our test analytics rule, and since it is enabled, it will run almost immediately and create our test alert and incident.

18. Since our rule will run every 5 minutes, and we don't need incidents to be created every 5 minutes, let's first go and disable our scheduled analytics rule. From Microsoft Sentinel, select the **Analytics** tab, select our **Test - Malicious IP** rule, and click on **Disable**, as shown in the following screenshot:

Figure 6.58 – Disable analytic rule

19. Now click on the **Incident** tab, and there should be one incident with the title **Test - Malicious IP**. To see quick details about the incident, select it, and details will appear on the right side, as shown in *Figure 6.59*:

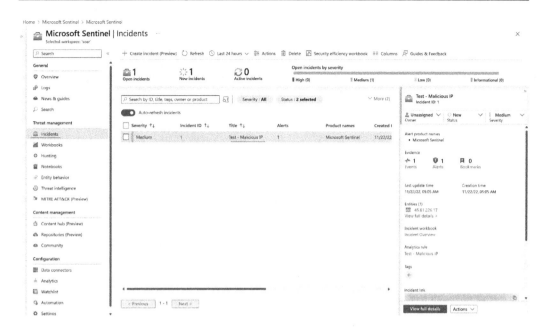

Figure 6.59 – Select incident created and click on View full details

20. Now, we can select **View full details** to see all the details of the incident.

Once opened, from the right side, we can see that our incident is already enriched from our playbook, as shown in *Figure 6.60*:

Figure 6.60 – Comment added to the incident overview page

21. Option number one is to run the playbook on an alert automatically on alert creation using automation rules. The exact process of automation rule creation can be done from the **Automation** tab, especially if we have multiple analytic rules that we want to include.

The second option is to run the playbook on an alert manually. To do that, we need to select the **Alert** tab from our incident overview page and scroll to the right until we see **View playbooks**, as shown in *Figure 6.61*:

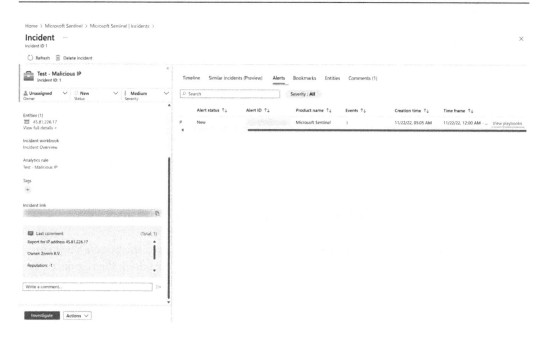

Figure 6.61 – Select the Alerts tab to run the alert playbook manually

22. By selecting **View playbooks**, we will see the list of playbooks that utilize the Microsoft Sentinel alert trigger. By selecting **Run**, we can run the playbook on the alert manually, as shown in the following screenshot:

Figure 6.62 – Select Run next to the playbook

23. We also have a **Runs** tab. That will show all playbook runs on this alert:

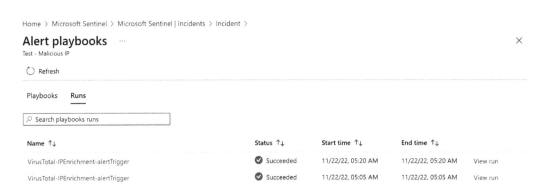

Figure 6.63 – Run history for this specific alert

24. If some of the playbooks failed, or we want playbook run details, we can click on **View run**, which will open the details, and we can investigate our playbook and check the responses, as shown in the following screenshot:

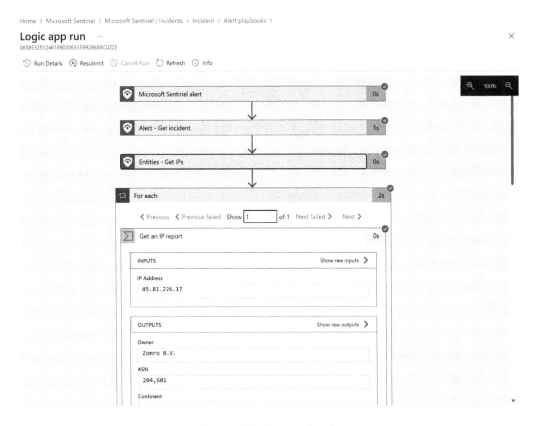

Figure 6.64 – See run details

25. When navigating through incident investigation and Microsoft Sentinel, always utilize the top menu, which will lead you back to the previous step or steps. In this case, we want to return to the incident, so we will select **Incident** in the menu:

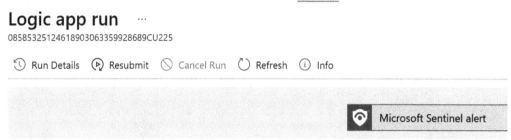

Figure 6.65 – How to navigate back in Microsoft Sentinel

26. Once we are back in the incident, under the **Comments** tab, we will see two comments now – one from the automation rule and one from the manual run, as shown in *Figure 6.66*:

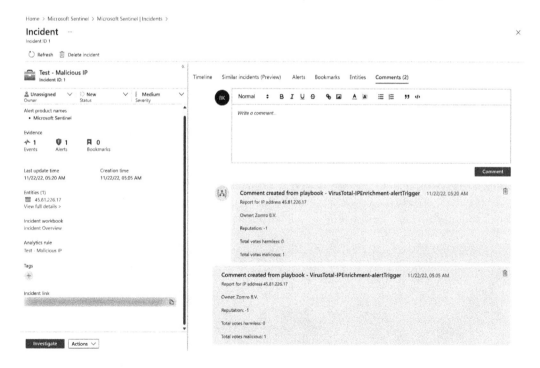

Figure 6.66 – See comments in the Comments tab on the incident overview page

Now, our SOC analyst has a better understanding of the IP address, and real benefits will be seen when we are speaking about multiple IP addresses and incidents that are regular occurrences in SOCs. In the next section, we will build a playbook that will utilize incident triggers to enrich URLs in incidents utilizing VirusTotal.

VirusTotal playbook – URL enrichment

In our second hands-on example, we will go through the process of creating a playbook that will utilize incident triggers, and we will run it on an incident that has an URL as an entity. This will help SOC analysts quickly get URL enrichment information and make faster decisions about how to proceed further with the investigation.

Before we begin, you will need the following:

- You need to have access to Microsoft Sentinel with appropriate permissions (Microsoft Sentinel Contributor, Logic App Contributor, and permission to assign RBAC controls – Owner or User Access Administrator).

- You need to have an authenticated VirusTotal Logic App connector (the one we created in the *VirusTotal playbook – IP enrichment* section). If you didn't perform the previous hands-on example, return to it for instructions on how to get the API key from VirusTotal.

Creating a playbook

In the first hands-on example, we used the alert trigger, but in this case, we will use an incident trigger. In Microsoft Sentinel, you will generally use an incident trigger in most cases, as an incident is what we are investigating, and an alert is seen as evidence of the incident. The alert trigger is mainly used for Microsoft Sentinel Scheduled analytic rules for which we don't want to create incidents (**Incident settings** in the analytic rule creation wizard) – for example, if there is an alert with no incident created but, using enrichment, we find that the URL or IP is malicious, and we want to create an incident using the playbook or send a notification to the SOC team. *Figure 6.67* shows how you can disable incident creation when creating or editing a scheduled rule:

Home > Microsoft Sentinel | Analytics >

Analytics rule wizard - Edit existing scheduled rule   ...
Test - Malicious IP

General   Set rule logic   **Incident settings**   Automated response   Review and update

Incident settings
Microsoft Sentinel alerts can be grouped together into an Incident that should be looked into.
You can set whether the alerts that are triggered by this analytics rule should generate incidents.

Create incidents from alerts triggered by this analytics rule

(**Enabled**   Disabled)

Alert grouping
Set how the alerts that are triggered by this analytics rule, are grouped into incidents.

Figure 6.67 – Disable incident creation when creating a new detection rule

Let's start with playbook creation:

1. To start exercise 2, go to Microsoft Sentinel, and under the **Automation** tab, select **Create** and
    then **Playbook with incident trigger**, as shown in the following screenshot:

Figure 6.68 – Create a playbook with an incident trigger

2. From the playbook creation wizard, under the **Basics** tab, we need to enter the playbook name (in my case, `VirusTotal-URLEnrichment-incidentTrigger`), and let's turn on the diagnostic settings and save to our demo Log Analytics workspace, as shown in the following screenshot:

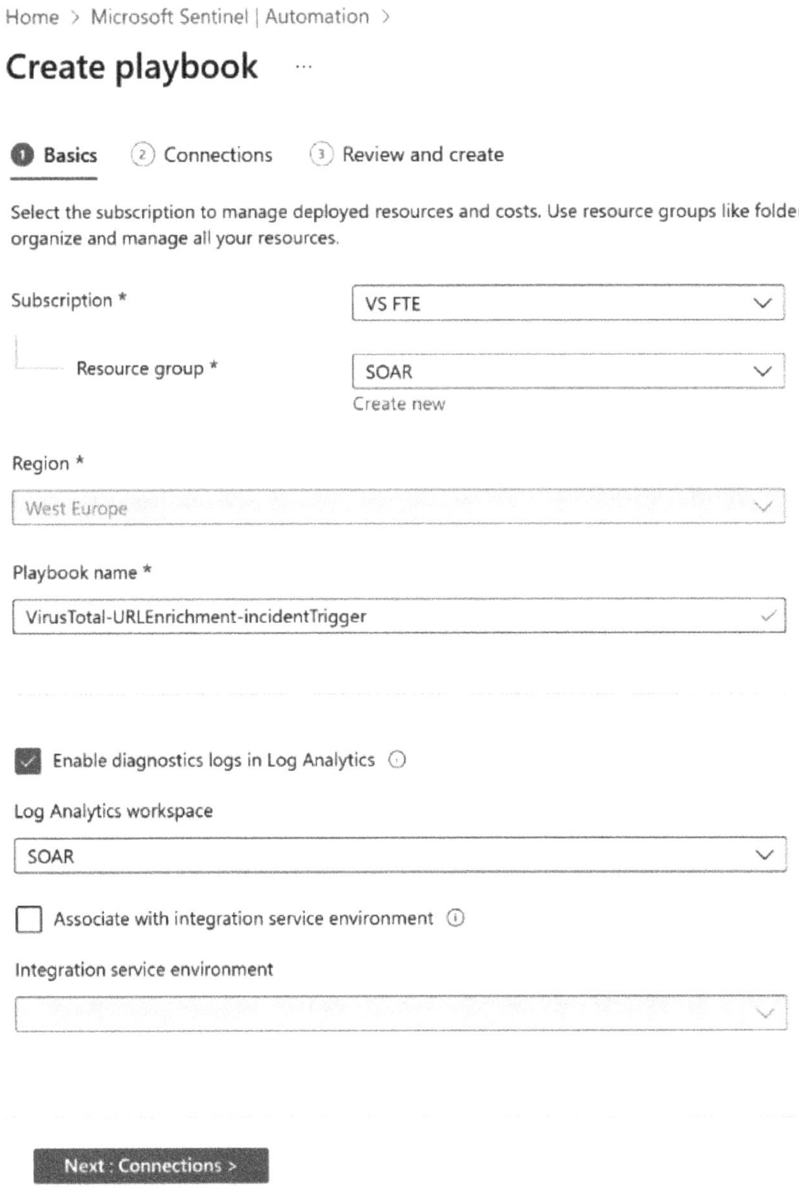

Home > Microsoft Sentinel | Automation >

Create playbook ...

1 **Basics**   ② Connections   ③ Review and create

Select the subscription to manage deployed resources and costs. Use resource groups like folders to organize and manage all your resources.

Subscription *

| VS FTE | ∨ |

Resource group *

| SOAR | ∨ |

Create new

Region *

| West Europe | ∨ |

Playbook name *

| VirusTotal-URLEnrichment-incidentTrigger | ✓ |

☑ Enable diagnostics logs in Log Analytics ⓘ

Log Analytics workspace

| SOAR | ∨ |

☐ Associate with integration service environment ⓘ

Integration service environment

| | ∨ |

Next : Connections >

Figure 6.69 – Enable diagnostic logs

3. The **Connections** tab remains unchanged as we use a managed identity. Under **Review and create**, check the configuration, select **Create**, and continue to the designer, as shown in *Figure 6.70*:

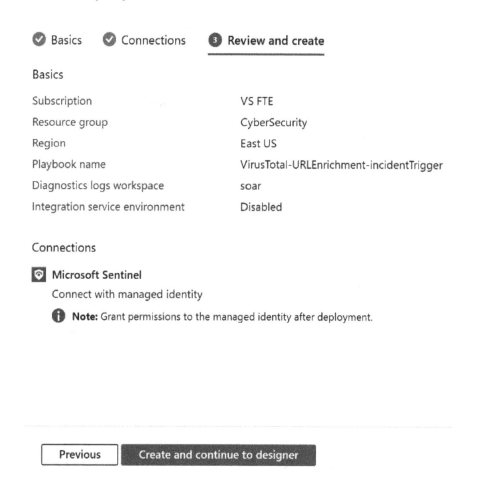

Home > Microsoft Sentinel | Automation >

Create playbook   ...

✅ Basics   ✅ Connections   ③ **Review and create**

Basics

Subscription	VS FTE
Resource group	CyberSecurity
Region	East US
Playbook name	VirusTotal-URLEnrichment-incidentTrigger
Diagnostics logs workspace	soar
Integration service environment	Disabled

Connections

🔷 **Microsoft Sentinel**
Connect with managed identity

ⓘ **Note:** Grant permissions to the managed identity after deployment.

Previous	Create and continue to designer

Figure 6.70 – Review and create new playbook

Something that I usually do is draw on a piece of paper how I want my playbook to look. This prepares me for thinking about requirements that are connected to the playbook.

This is my thought process for this playbook:

1. We will use an incident trigger (we have done this) and I will utilize a managed identity for Microsoft Sentinel triggers and actions (managed identity is enabled and is used automatically in the playbook – step 2 of the playbook creation wizard, **Connections**).

2. We will need to extract URLs from the **Entity** array, and we will utilize the **Entity - Get URLs** action. We will utilize a managed identity, and no additional permissions are needed, as we are extracting information that is already in the trigger.

3. Next, we will need to add a VirusTotal action for URL enrichment. We need to authenticate the VirusTotal Logic App connector. We did this in the first exercise.

4. I want to add a comment to the incident with enrichment info for each URL enrichment. From the first exercise, we know that to add a comment to the incident, we will need an incident ARM ID, and we didn't have it with the alert trigger. In the incident trigger, we have it, so we will not need an action to grab incident details.

   Since we are adding comments to Microsoft Sentinel, do we need to assign permission to our managed identity? As this is an update on the incident, we will need to assign the **Microsoft Sentinel Responder** role to the managed identity.

Now, this is easy for me as I work with Microsoft Sentinel automation daily. But how can you get to that stage? For this, I started step by step. I don't create a playbook first, but I create the incident that I can use to test the playbook. So let's create one for URL enrichment.

We will utilize watchlists again to add the URL that we will use. The process is the same as in the first exercise, but we will use URL instead of IP:

1. First, let's open a blank Microsoft Excel workbook. In cell A1, type URL, and in cell A2, insert www.bcomb.net (or any other URL of your choosing).

   Save it as a CSV file. (I used URL watchlist.csv.)

2. In Microsoft Sentinel, go to the **Watchlist** tab and select **Add new**. In **General**, enter MaliciousURL as the name and alias.

   In the **Source** tab, upload our URL watchlist and select **URL** for **SearchKey**, as shown in the following screenshot:

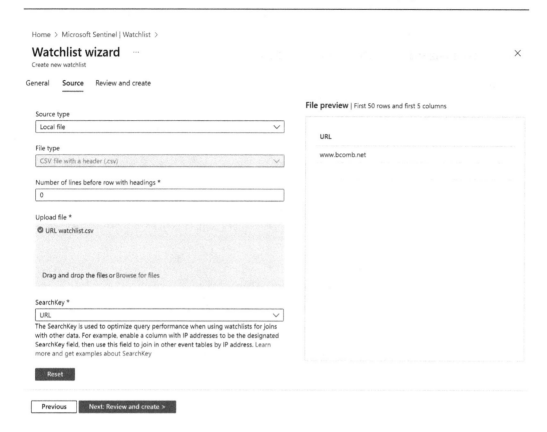

Figure 6.71 – The same view as with the IP watchlist; update the SearchKey value

3. Select **Next: Review and create** and then **Create**.

    The same as for the IP watchlist, we will need to wait 5-10 minutes until we can see the data in the logs.

4. Once the data is uploaded (we can see the number of rows has increased to **1**), select **View in logs.**

5. Now, create the Microsoft Sentinel schedule rule from it by selecting **New alert rule** and **Create Azure Sentinel alert**.

6. It will open the analytics rule creation wizard. I named it `Test - Malicious URL`, and I left the rest unchanged. Click on **Next: Set rule logic**.

    In **Set rule logic**, we will first set **Entity Mapping** by selecting **URL** as the entity type, **Url** as the identifier, and **URL** as the value, as shown in the following screenshot:

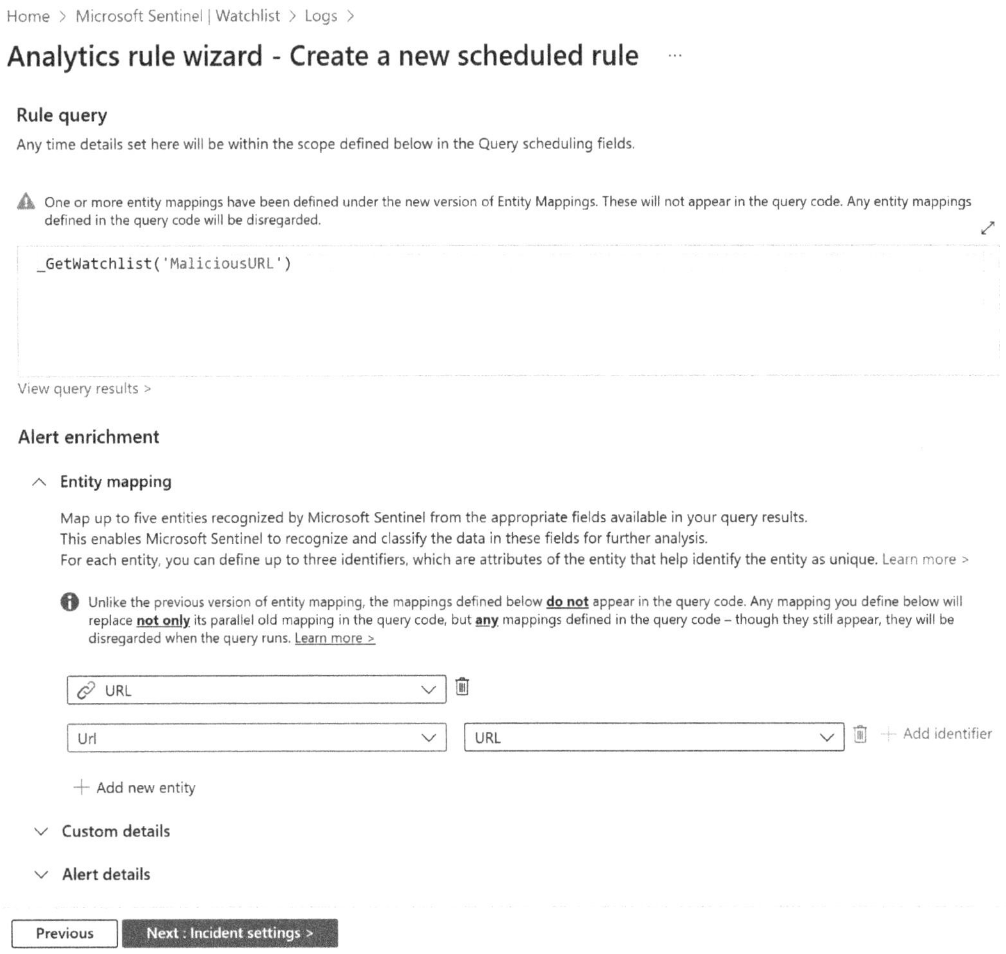

Home > Microsoft Sentinel | Watchlist > Logs >

Analytics rule wizard - Create a new scheduled rule   ...

Rule query

Any time details set here will be within the scope defined below in the Query scheduling fields.

⚠️ One or more entity mappings have been defined under the new version of Entity Mappings. These will not appear in the query code. Any entity mappings defined in the query code will be disregarded.

```
_GetWatchlist('MaliciousURL')
```

View query results >

Alert enrichment

⌄ **Entity mapping**

Map up to five entities recognized by Microsoft Sentinel from the appropriate fields available in your query results.
This enables Microsoft Sentinel to recognize and classify the data in these fields for further analysis.
For each entity, you can define up to three identifiers, which are attributes of the entity that help identify the entity as unique. Learn more >

ℹ️ Unlike the previous version of entity mapping, the mappings defined below **do not** appear in the query code. Any mapping you define below will replace **not only** its parallel old mapping in the query code, but **any** mappings defined in the query code – though they still appear, they will be disregarded when the query runs. Learn more >

| 🔗 URL | ⌄ | 🗑️ |

| Url | ⌄ | URL | ⌄ | 🗑️ + Add identifier |

+ Add new entity

⌄ **Custom details**

⌄ **Alert details**

[Previous] [**Next : Incident settings >**]

Figure 6.72 – Entity mapping for URL value

7. The last step in **Set rule logic** here will be to change **Run query every** from **Hours** to **Minutes**, and then we can click on **Next: Incident settings**.

   We will leave the rest of the tabs unchanged, and in **Review and create**, we can check our configuration and click on **Create**, as shown in the following screenshot:

Analytics rule wizard - Create a new scheduled rule   ...

✓ Validation passed.

General   Set rule logic   Incident settings   Automated response   **Review and create**

Analytics rule details

Name	Test – Malicious URL
Description	
Tactics and techniques	
Severity	Medium
Status	Enabled

Analytics rule settings

Rule query	_GetWatchlist('MaliciousURL')
Rule frequency	Run query every **5 minutes**
Rule period	Last **5 hours** data
Rule start time	Automatic
Rule threshold	Trigger alert if query returns **more than 0** results
Event grouping	Group all events into a single alert
Suppression	Not configured

Entity mapping

Entity 1:	**URL** Identifier: Url, Value: URL

Previous   Create

Figure 6.73 – Review the configuration and save the analytics rule

In this case, we didn't attach the automation rule immediately as our playbook has still not been created.

8. Since our rule will run every 5 minutes, and we don't need incidents to be created every 5 minutes, let's first go and disable our scheduled analytics rule. From Microsoft Sentinel, select the **Analytics** tab, then select the **Test -Malicious URL** rule, and click on **Disable**.

9. Now click on the **Incidents** tab, and there should be one incident with the title **Test - Malicious URL**. To see quick details about the incident, select it, and details will appear on the right side. Click on **Actions**, and select **Run playbook**, as shown in the following screenshot:

Figure 6.74 – Run the playbook manually on the incident from the Actions submenu

10. This will open the **Run playbook on incident** wizard, where we can see all our incident trigger playbooks. We can mark any as our favorite so they always appear at the top. Find **VirusTotal-URLEnrichment-incidentTrigger** and select **Run**, as shown in the following screenshot:

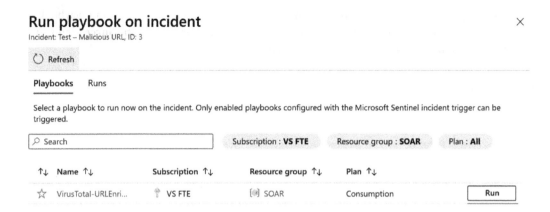

Figure 6.75 – Select Run next to our incident playbook

11. What we have done now is run our playbook with only a trigger. If you are starting out with Microsoft Sentinel playbooks, this can show you what kind of information the incident trigger returns in the playbook. From the **Run playbook on incident** wizard, click on the playbook name to open it, as shown in the following screenshot:

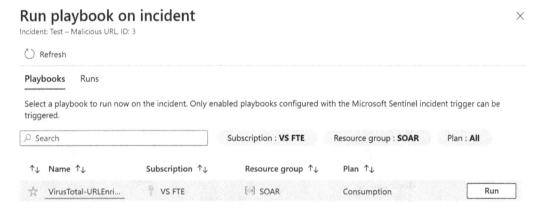

Figure 6.76 – Open the playbook from the Run playbook on incident wizard

12. This will open the **Overview** tab of our playbook, where we can see the details of our playbook, including playbook runs, which are located under **Runs history**. Select the **Succeeded** run from the view, as shown in the following screenshot:

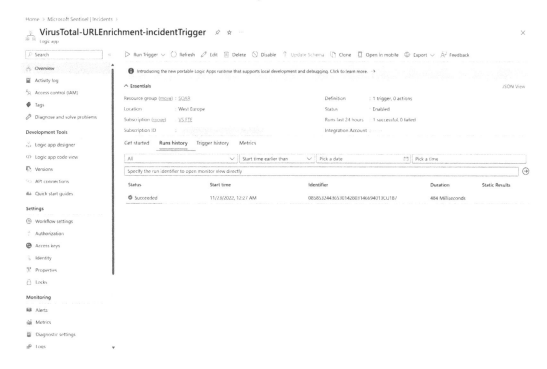

Figure 6.77 – Select Succeeded under Runs history

13. This will open details about the run, as shown in the following screenshot:

Figure 6.78 – Runs history view

14. Let's select **Microsoft Sentinel incident** from **Logic app run** on the details page.

In the output, we will see all the details our trigger contains about the incident, which we can utilize in our playbook. That data, for example, is severity, status, details of each entity, details about each alert, tags, comments, and so on. We can use this data in our playbooks with dynamic content or expressions, as shown in the following screenshot:

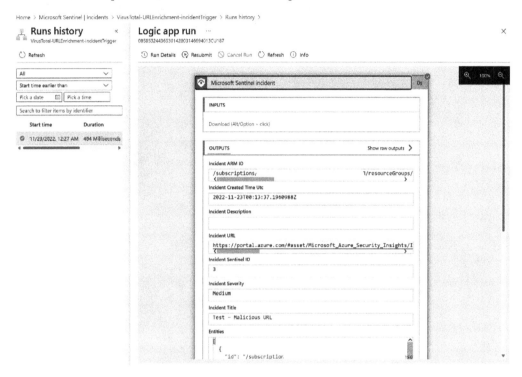

Figure 6.79 – Expand trigger and or action for details

15. If we click on **Show raw outputs**, we see the whole response in JSON format, as shown in the following screenshot:

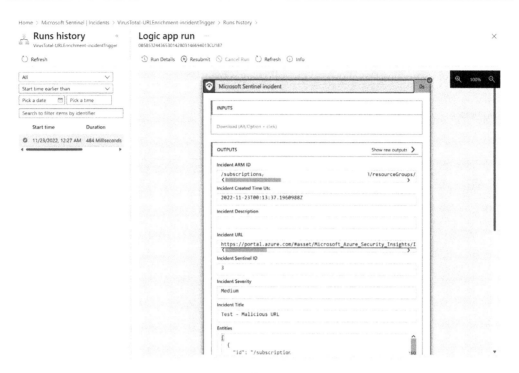

Figure 6.80 – Show raw outputs

Figure 6.81 shows us the JSON of the raw output:

```
"headers": {
    "Accept-Encoding": "gzip,deflate",
    "Host": "prod-254.westeurope.logic.azure.com",
    "x-ms-client-tracking-id": "e6c70732-b8bc-46ff-998e-42e3af957b5d_3",
    "x-ms-correlation-request-id": "0b8aabe1-ae6a-439b-bbf4-2af41fbb3304",
    "x-ms-forward-internal-correlation-id": "f647ead3-5cd8-4574-ab18-3700680386af",
    "Content-Length": "4110",
    "Content-Type": "application/json; charset-utf-8"
},
"body": {
    "eventUniqueId": "62222d61-873a-47ed-b4d5-b35cd23eb5b1",
    "objectSchemaType": "Incident",
    "objectEventType": "Create",
    "workspaceInfo": {
        "SubscriptionId": "8031140b9-666a-4e5c-957c-cff7b5029278",
        "ResourceGroupName": "soar",
        "WorkspaceName": "soar"
    },
    "workspaceId": "e6c70732-b8bc-46ff-998e-42e3af957b5d",
    "object": {
        "id": "/subscriptions/                  /resourceGroup   /providers/Microsoft.OperationalInsights/workspaces/   /providers/Microsoft.SecurityInsights/Incidents/b
        "name": "hdaf76be-d265-4186-a08e-94e99f610e1a",
        "etag": "\"7701f6e0-0000-0d00-0000-637d65b10000\"",
        "type": "Microsoft.SecurityInsights/Incidents",
        "properties": {
            "title": "Test - Malicious URL",
            "description": "",
            "severity": "Medium",
            "status": "New",
            "owner": {
                "objectId": null,
                "email": null,
                "assignedTo": null,
                "userPrincipalName": null
            },
            "labels": [],
            "firstActivityTimeUtc": "2022-11-22T19:08:35.655Z",
            "lastActivityTimeUtc": "2022-11-23T00:08:35.655Z",
            "lastModifiedTimeUtc": "2022-11-23T00:13:37.1960988Z",
```

Figure 6.81 – Raw output in JSON format

This will be useful once you start working with playbooks in more depth as you can find and expose data that may not be available with dynamic content or when using an HTTP action and you need to parse a JSON response.

16. Okay, let's get back to our playbook, as shown in the following screenshot:

Home > Microsoft Sentinel | Incidents > VirusTotal-URLEnrichment-incidentTrigger > Runs history > Logic app run >

Outputs ···
Microsoft Sentinel incident

```
{
    "headers": {
        "Accept-Encoding": "gzip,deflate",
        "Host": "prod-254.westeurope.logic.azure.com",
        "x-ms-client-tracking-id": "e6c70732-b8bc-46ff-998e-42e3af957b5d_3",
        "x-ms-correlation-request-id": "0b8aabe1-ae6a-439b-bbf4-2af41fbb3304",
        "x-ms-forward-internal-correlation-id": "f647ead3-5cd8-4574-ab18-37006803b6af",
```

Figure 6.82 – Navigate to the playbook overview tab

And from the right menu, select **Logic app designer**, as shown in the following screenshot:

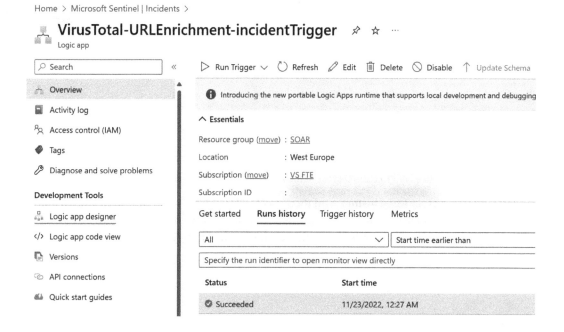

Figure 6.83 – Access Logic app designer

So, we have added the incident trigger, the first step in the playbook creation process, and we saw the response of the playbook run. We need to proceed to the next step of our playbook creation process, which is to extract the URL entity from the incident using the **Entity - Get URLs** action:

1. Click on **New step**, select **Microsoft Sentinel** from the recent connections (or search for `Microsoft Sentinel`), and select the **Entities - Get URLs** action. Add **Entities** from **Dynamic content**.

2. If we want to see each dynamic field of **Entities - Get URLs**, so we know what to use, we can rerun the playbook. First, save the playbook by selecting **Save**.

    An easy way to do it is to open the playbook in a new tab so that you can perform testing and creation at the same time. To do so, open a new tab, go to **Microsoft Sentinel**, then **Automation | Active playbooks**, and find and open our playbook (**VirusTotal- URLEnrichment-incidentTrigger**). Or, duplicate the tab you are currently working on.

3. Once it is open, select the last run from the run history. From the **Logic app run** window, select **Resubmit**, as shown in the following screenshot:

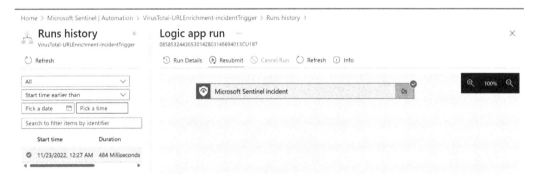

Figure 6.84 – Resubmit playbook run

This will rerun our playbook and use the same incident details that were used in the previous run.

4. Select **Refresh** from **Runs history**, and then select the last run. From **Logic app run**, expand the **Entities - Get URLs** action by clicking on it, and scroll down until we see the body of our output. We will see what each field is in the scheme, as shown in *Figure 6.85*:

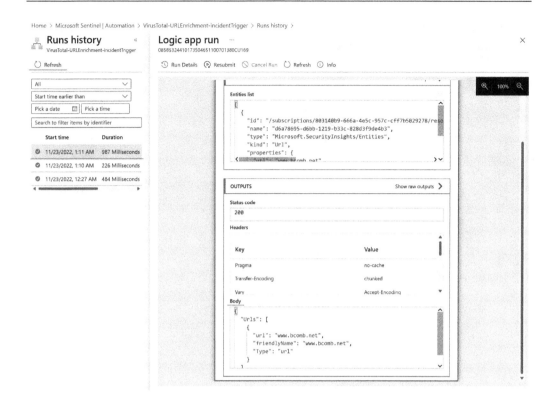

Figure 6.85 – Refresh to access new details

This is how we can resubmit the run after each step when we need the playbook's performance and data before continuing to the next step. It's important to note that you first need to save the playbook before resubmitting it (or you will have the same run as the previous one, as you can see in the middle run in the preceding screenshot).

Now let's go back to the browser tab with our playbook design, and it's time for the next step of the playbook creation process. In this step, we need to utilize a VirusTotal action to check the URL reputation to enrich incidents with it:

1. We will select **New step**, search for `VirusTotal`, and select the **Analyze a URL** action. We will add **URLS Url** from **Dynamic content**, and the same as with IPs in the first exercise, it will create a **For each** loop so that we can analyze all URLs from the incident, as there can be more than one.

2. We will save the playbook again and resubmit it to get information from VirusTotal and to see what data each input contains and what will be helpful for SOC analysts to perform faster incident triage.

We will see that the output is not helpful for our use case and that we need to change something, as shown in the following screenshot:

Figure 6.86 – Analyze a URL response

3. Now, we will go to the playbook design, click on **Add an action** in the **For each** loop, search for `VirusTotal`, and check other actions. One of them will be **Retrieve information about a file or URL analysis**. We will select it, and as input, we will use the output of the **Analyze a URL** action using **Dynamic content**. Click **Save** again. This is shown in the following screenshot:

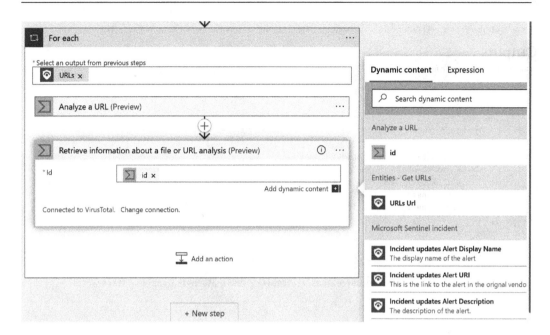

Figure 6.87 – Retrieve information about a file or URL analysis action

4. Let's again go to the browser tab where we have run the playbook earlier, resubmit the playbook, and check the **Retrieve information about a file or URL analysis** action output. Now we will see more data, and if we select **Show raw output**, we will see all the details, including the status of how many reported it harmless, how many reported it malicious, and so on. This is shown in the following screenshot:

Home > Microsoft Sentinel | Automation > VirusTotal-URLEnrichment-incidentTrigger > Runs history > Logic app run >

Outputs   ...

Retrieve information about a file or URL analysis

```
{
    "statusCode": 200,
    "headers": {
        "X-Cloud-Trace-Context": "64690eab866b59865a970da31b0181dc",
        "Timing-Allow-Origin": "*",
        "x-ms-apihub-cached-response": "true",
        "x-ms-apihub-obo": "false",
        "Date": "Wed, 23 Nov 2022 01:44:24 GMT",
        "Content-Length": "21698",
        "Content-Type": "application/json"
    },
    "body": {
        "meta": {
            "url_info": {
                "url": "http://bcomb.net/",
                "id": "efa31ce0c46fc0238d3c91e529946029599bada55fb525d2ed3cd34b3f6447e7"
            }
        },
        "data": {
            "attributes": {
                "date": 1669167505,
                "status": "completed",
                "stats": {
                    "harmless": 65,
                    "malicious": 15,
                    "suspicious": 0,
                    "undetected": 11,
                    "timeout": 0
                },
                "results": {
                    "Bkav": {
                        "category": "undetected",
                        "result": "unrated",
                        "method": "blacklist",
                        "engine_name": "Bkav"
                    },
                    "CMC Threat Intelligence": {
                        "category": "harmless",
                        "result": "clean",
                        "method": "blacklist",
```

Figure 6.88 – Retrieve information about a file or URL analysis raw details

This looks much better, and we can now continue to the fourth step of the playbook creation process, which is to add helpful information to the incident comment:

1. Let's select **Add an action | Microsoft Sentinel** and select the **Add comment to incident** action. First, we will add the incident ARM ID to identify the incident where the comment should be added, as shown in the following screenshot:

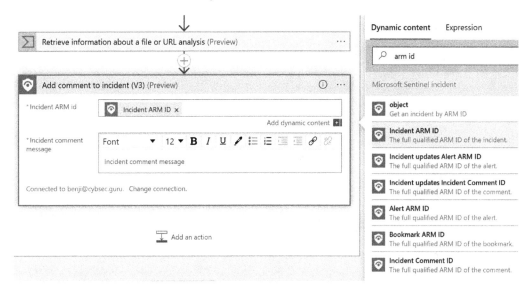

Figure 6.89 – Incident ARM ID available with the use of the incident trigger

2. Then we need to write the body of a comment:

   I. First, we will add the following text:

   • VirusTotal analysis for URL

   • Harmless:

   • Suspicious:

   • Malicious:

   • Undetected:

   II. And then, we will add **Dynamic content** from our **Retrieve information about a file or URL analysis** action, as shown in the following screenshot:

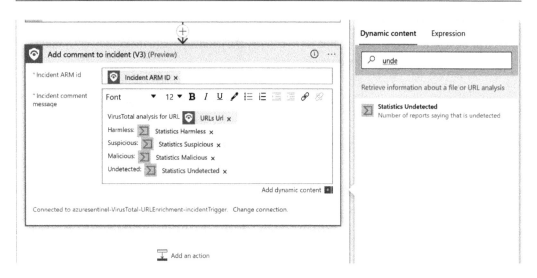

Figure 6.90 – Fill in incident comment message details

3. Now we can save our playbook. Make sure that Microsoft Sentinel is using managed identity for authentication. This will be visible at the bottom of the action, where it shouldn't be the user's email address but the text `azuresentinel-VirusTotal-URLEnrichment-incidentTrigger`. If it's a user identity, click on **Change connection** and select a connection from the list. There is no need to click on **Add new**, as we already created one when creating the playbook. This is shown in the following screenshot:

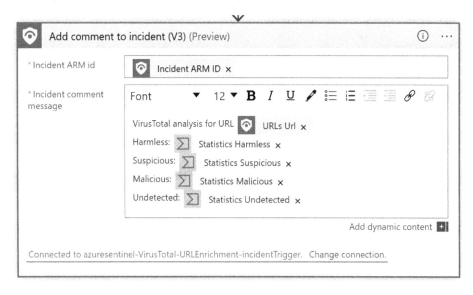

Figure 6.91 – Check that managed identity is being used

Figure 6.92 shows us the managed identity we need to select:

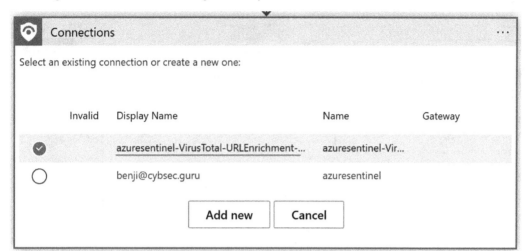

Figure 6.92 – Select managed identity from the list if a user identity is used

With this, we have created our playbook. What we need to do next is to test and see whether we missed something or whether we need to change certain elements in the playbook.

Testing a playbook

Once we have created everything, we can test our playbook:

1. To test it, we can use **Resubmit** again. One thing that we will notice is that the playbook has failed. If we expand the failed playbook run, we will see that it failed on the **Add comment to incident** action. By selecting the **Add comment to incident** action, it will expand, and we will see a **Forbidden** notification, and in **OUTPUTS**, we will see a message that the authorization failed, as shown in the following screenshot:

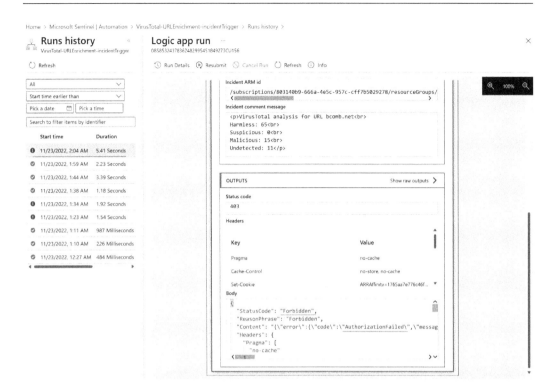

Figure 6.93 – Authorization failed message

2. This is expected as we need to add a comment to the incident and use a managed identity that doesn't have Microsoft Sentinel permissions. What we need to do is to go back to our playbook design and go to the playbook overview. From the left menu, we will select **Identity**. We will see that status is **On**, which means that managed identity is enabled. What we need to do is to select **Azure role assignments** to assign the role, as shown in the following screenshot:

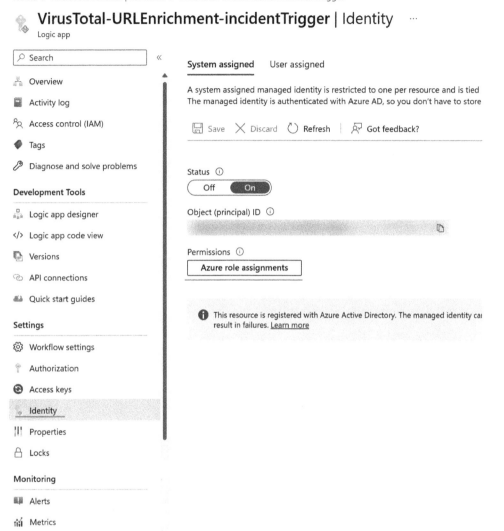

Figure 6.94 – Access the managed identity tab

3. Next, we need to click on **Add role assignment**, select our scope (in my case, **Resource group**), and add the role. In this case, we need **Microsoft Sentinel Responder**. Click **Save**, as shown in *Figure 6.95*:

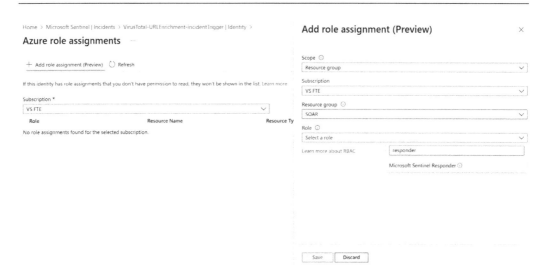

Figure 6.95 – Assign needed permission

4. Let's go back to **Runs history** and resubmit the run. Now we will see that the playbook run is successful, as shown in the following screenshot:

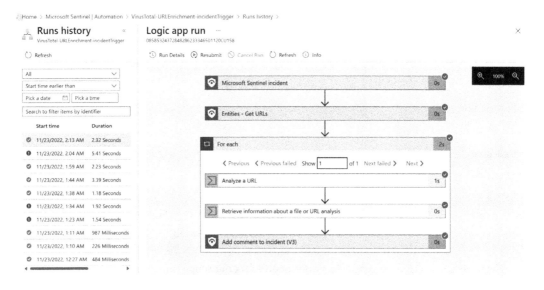

Figure 6.96 – Resubmit and check the details

5. We can go back to our incident in Microsoft Sentinel by selecting the **Incidents** tab and selecting the incident (**Test – Malicious URL**). In the right pane, if we scroll down, we can see the comment added by our playbook, as shown in the following screenshot:

Figure 6.97 – View comment from the Incidents pane after selecting the incident

Now, a SOC analyst can quickly get this info for triage, and you can work with the SOC analyst to see whether any other helpful information can be added from the VirusTotal response.

6. To manually run the playbook on the incident, we can do it from the following places:

    •   **Option one**: The **Incidents** page – Left-click on the incident and select **Run playbook**, as shown in the following screenshot:

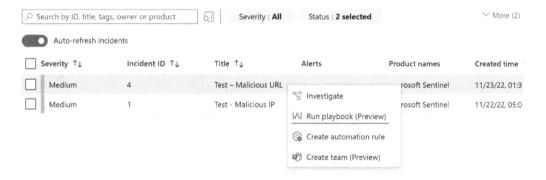

Figure 6.98 – Manually run the playbook on the incident – option one

- **Option two**: The **Incidents** page – From the details page, select **Actions** and then **Run playbook**, as shown in the following screenshot:

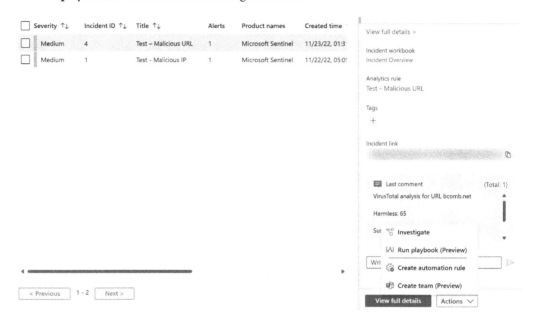

Figure 6.99 – Manually run the playbook on the incident – option two

- **Option three**: The **Incident** overview page – Select **Actions** and then **Run playbook**, as shown in the following screenshot:

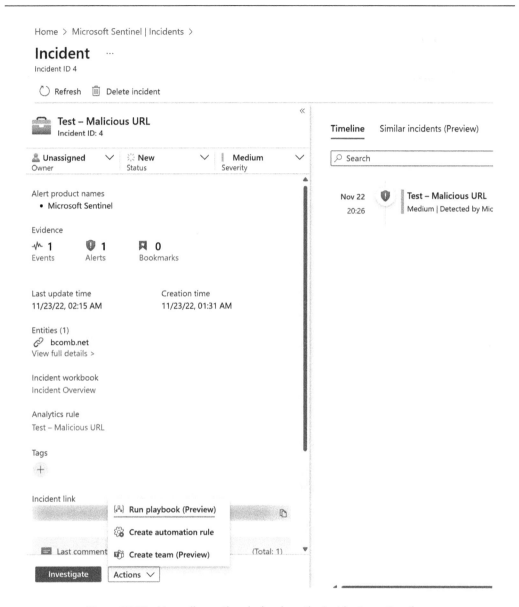

Figure 6.100 – Manually run the playbook on the incident – option three

- **Option four**: To run the playbook automatically on incident creation, we will need to add it to the automation rule by using the **When incident is created** trigger and **Run playbook** as the action:

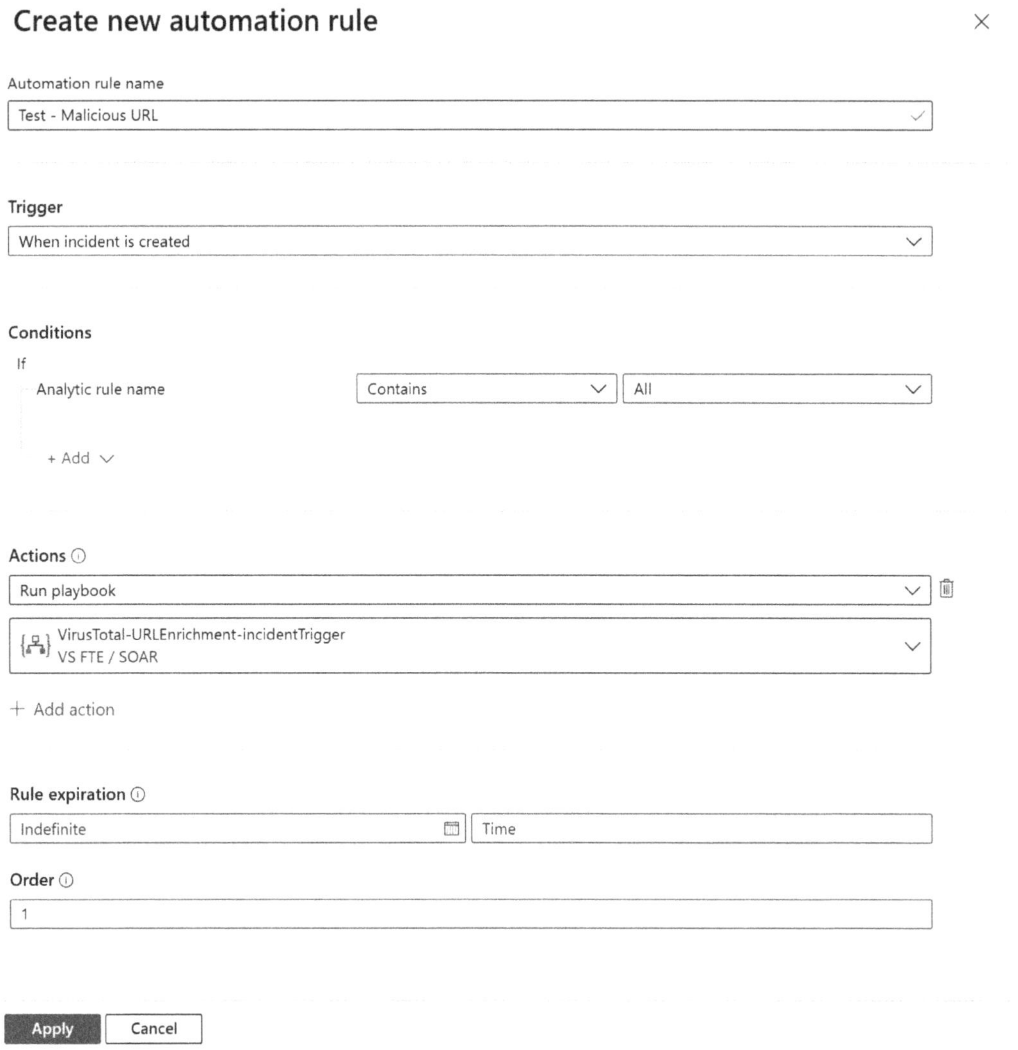

Create new automation rule   ✕

Automation rule name

```
Test - Malicious URL                                                    ✓
```

Trigger

```
When incident is created                                                ⌄
```

Conditions

If

| Analytic rule name | Contains ⌄ | All ⌄ |

+ Add ⌄

Actions ⓘ

```
Run playbook                                                        ⌄   🗑
```

| {🔧} VirusTotal-URLEnrichment-incidentTrigger
 VS FTE / SOAR | ⌄ |

+ Add action

Rule expiration ⓘ

```
Indefinite                                        📅   Time
```

Order ⓘ

```
1
```

Apply   Cancel

Figure 6.101 – Run the playbook automatically on incident creation

This is how we can run our newly created playbook. But we can also monitor the run details of the playbook as we have turned on diagnostic settings for the playbook. Run this query in **Microsoft Sentinel Logs** to see details:

```
AzureDiagnostics
| where OperationName == "Microsoft.Logic/workflows/
workflowRunCompleted"
```

Figure 6.102 shows us an example of the query run with details around the playbook run:

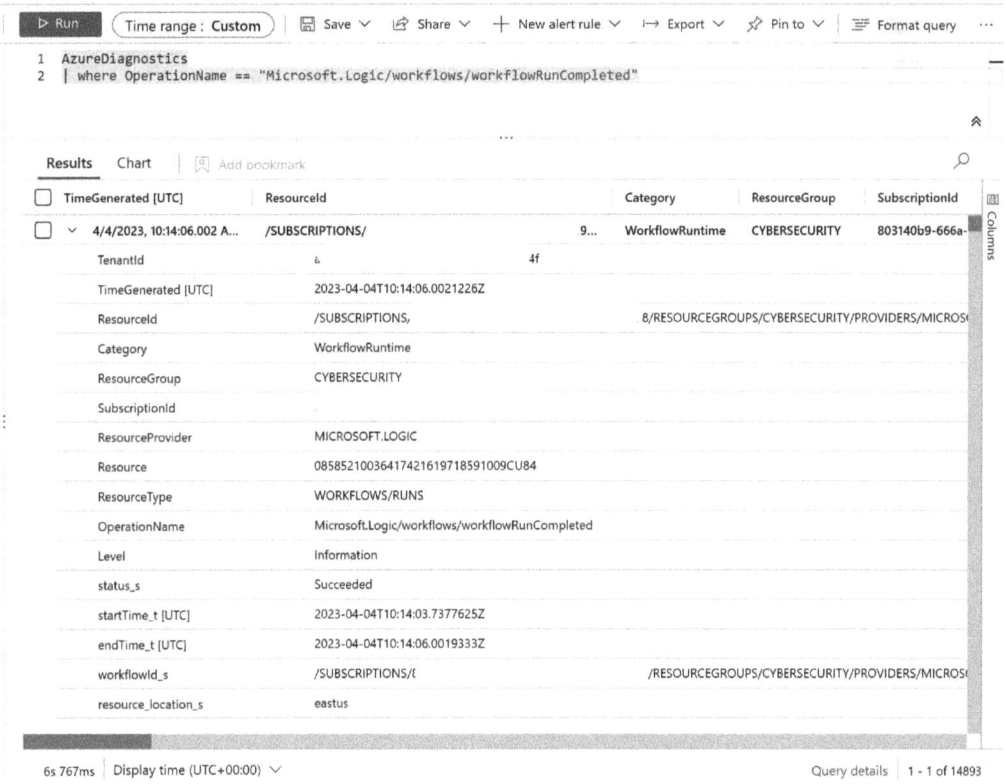

Figure 6.102 – Playbook diagnostic details

After checking the run details of our playbook, we will conclude the examples in this chapter.

Summary

In this chapter, we started with our hands-on examples. We focused on the enrichment of incidents so that we could speed up MTTA and MTTR in Microsoft Sentinel. We introduced how to access a trial of Microsoft Sentinel, continued with step-by-step playbook creation, and finished with an example of how to test our playbooks. We used two enrichment methods – IP and URL enrichment – and two different playbook triggers – alert and incident triggers.

In our second hands-on example, we went through the step-by-step process of creating a playbook while investigating each action and its output to plan what to utilize in the next action. This is important, especially when dealing with new actions, and we need to figure out what kind of information each piece of dynamic content contains.

In the next chapter, we will focus more on incident management automation, which will involve the expansion of the playbooks we created in the first and second examples, and making more automated responses based on watchlist data, as well as user input using email, Microsoft Teams chat, or channel messages.

7

Managing Incidents with Automation

In the previous chapter, we introduced how to use Microsoft Sentinel and then moved on to hands-on examples.

The first hands-on example involved enriching an incident that contained an IP address with information from VirusTotal. We used an alert trigger and went step by step from there, from creating a playbook to testing it.

The second example used URL enrichment, and we used a different approach to the IP enrichment example. We used an incident trigger for our playbook and went step by step through the usual process of creating the playbook.

This chapter will focus on how to manage incidents by utilizing automation.

This chapter will cover the following:

- Auto-closing known false-positive incidents using a watchlist
- Closing an incident based on SOC analyst input
- Auto-closing incidents using automation rules

Automated false-positive incident closure with a watchlist

Before we begin, you will need the following:

- You need to have access to Microsoft Sentinel with appropriate permissions (Microsoft Sentinel Contributor, Logic App Contributor, and permission to assign RBAC controls – Owner or User Access Administrator)

Creating a playbook

In this example, we will auto-close an incident automatically when the incident is created and when the IP address is an approved internal IP in a watchlist.

Let's use the same strategy we used with our hands-on example in the previous chapter. First, we will list what we want to do and then do it step by step:

1. We need a watchlist that contains an IP address. We have one called `MaliciousIP`, created in *Exercise 1* in the previous chapter. You should create a new watchlist called `AllowedIP` and use the same IP.

2. We will need a detection rule with an IP address. We created one in *Exercise 1* in the previous chapter called `Test - Malicious IP`. We will use this one again.

3. We need to create a playbook:

    I.   **Step A**: We will run it on the incident, so we need to utilize the incident trigger.

    II.   **Step B**: We need to extract IPs from the **Entities** array.

    III.   **Step C**: Get IPs from the watchlist using **Azure Monitor Logs** and compare them to IPs from the incident. We will create a service principal and assign it the **Log Analytics Reader** role.

    IV.   **Step D**: Compare an IP from the watchlist with the IP from the incident.

    V.   **Step E**: If the IP is on the `AllowedIP` watchlist, auto-close the incident and add the tag.

    VI.   **Step F**: Otherwise, add a comment to the incident, stating that the IP is not on the `AllowedIP` watchlist.

    VII.   **Step G**: We will use a managed identity, and we will need the **Microsoft Sentinel Responder** role to be able to auto-close incidents and add comments.

Let's start!

The first step involves creating a watchlist, and I have created a new one with the same IP (`45.81.226.17`); I just used a different name and alias – `AllowedIP`. We did this in the previous chapter, where you can find all the instructions.

The second step involves having a detection rule, and here I will use the one from the previous chapter, `Test - Malicious IP`.

The third and final step involves creating a playbook.

Step A – initializing a playbook and adding a trigger

The process of creating a playbook starts by initializing the playbook and adding a trigger:

1. Go to our familiar Microsoft Sentinel environment, then the **Automation** tab, select **Create**, and then **Blank playbook**.

2. The blank playbook creation wizard looks a little different, as we can now select whether we want to create a playbook using **Logic App Consumption** or **Logic App Standard**.

   We will select our subscription, resource group, and unique name and make sure we select **Standard** under **Plan** as shown in the following screenshot:

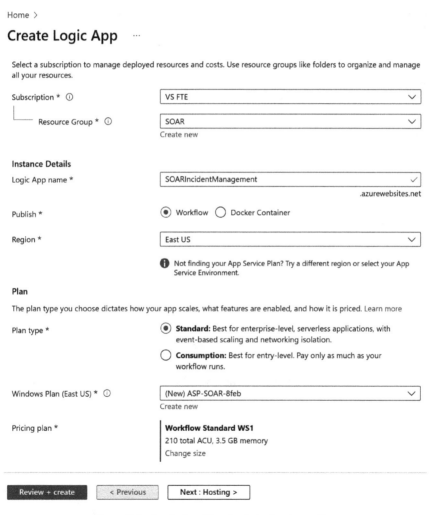

Figure 7.1 – Logic App Standard playbook creation

3. Click on **Next : Hosting >**. On the **Hosting** tab, we need to create or link to external storage that stores the necessary data, as shown in *Figure 7.2*:

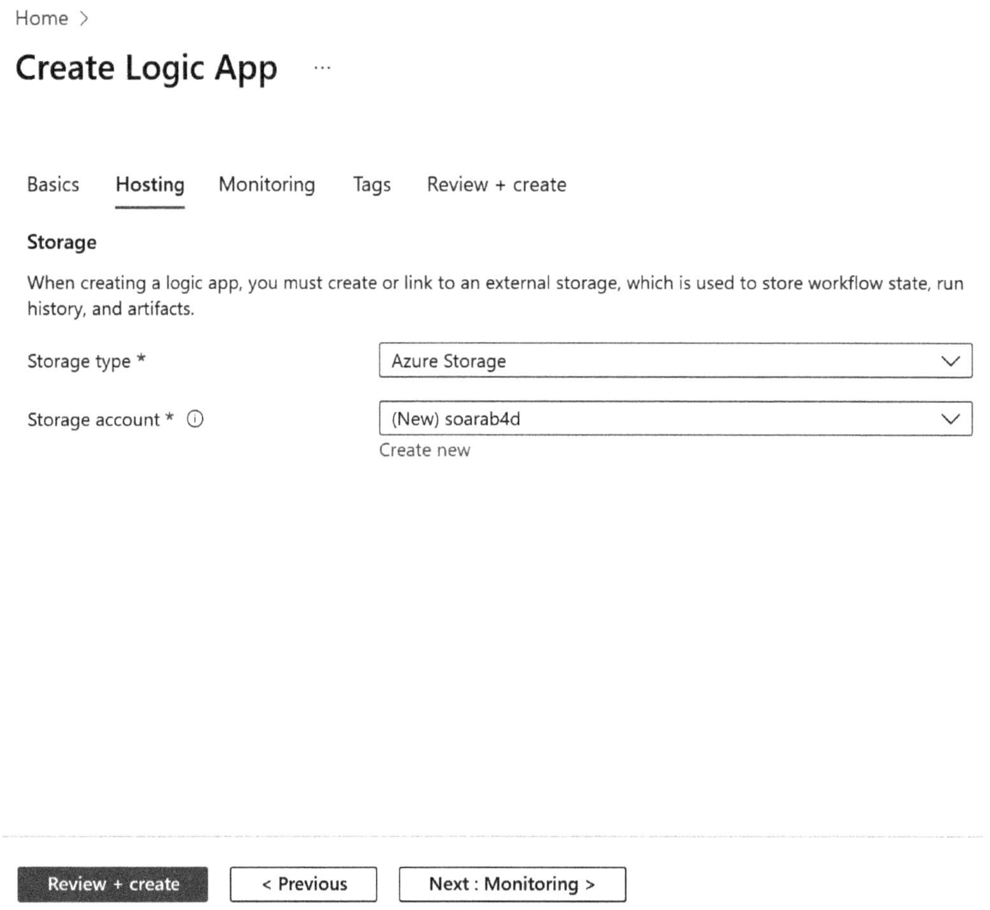

Home >

Create Logic App   ⋯

Basics   **Hosting**   Monitoring   Tags   Review + create

Storage

When creating a logic app, you must create or link to an external storage, which is used to store workflow state, run history, and artifacts.

Storage type *   Azure Storage   ∨

Storage account * ⓘ   (New) soarab4d   ∨
                            Create new

Review + create   < Previous   Next : Monitoring >

Figure 7.2 – Logic App Standard Hosting tab

4. Click on **Next** for **Monitoring** and **Tags**, and under **Review + create**, check your configuration and select **Create**.

Once the playbook is deployed, select **Go to resource**, as shown in the following screenshot:

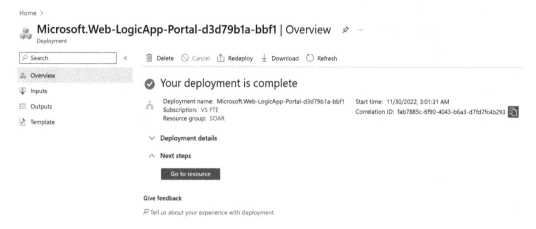

Figure 7.3 – Go to the newly created Logic App Standard resource

5.   From the left menu, select **Workflows**, as shown in *Figure 7.4*:

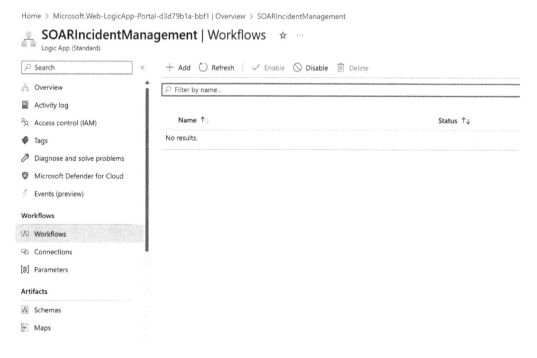

Figure 7.4 – Workflows in Logic App Standard

6. In **Workflows**, we will create our playbooks, and each playbook in a **Logic App Standard** will be an individual workflow.

    So, let's go and click on **Add**, enter a workflow name (that is, a playbook name; I used `AutoClose-IP-In-Watchlist`), and select a state. Please note that Microsoft Sentinel only supports the **Stateful** state. When you are done, select **Create**, as shown in *Figure 7.5*:

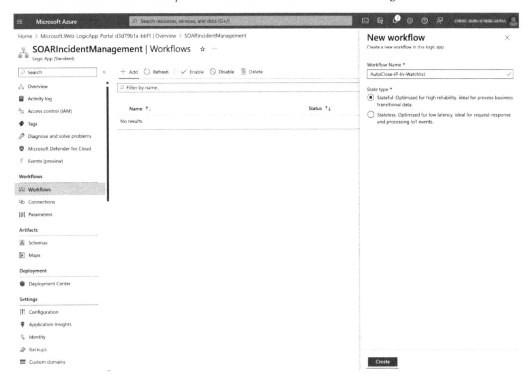

Figure 7.5 – Adding a new workflow to Logic App Standard

7. Once the workflow is created, we will see it on the list. Select it. The first page will show us an overview page similar to playbooks with **Logic App Consumption**, where we can see run and trigger history, as well as options to enable or disable a workflow, delete a workflow, and so on.

8. From the left menu, select **Designer**. The design approach remains unchanged. It will still contain triggers, actions, and dynamic content.

    Let's add a trigger. We will use the **Microsoft Sentinel incident** trigger. You will find it under **Azure**.

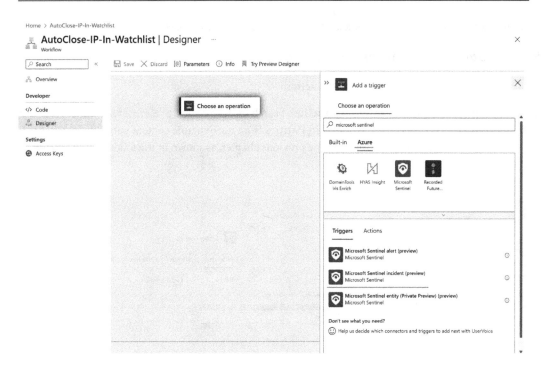

Figure 7.6 – Adding a trigger in the workflow

9. In **Create Connection**, select **Connect with a managed identity**. Enter the name and click on **Create**. Please note that this managed identity is on the **Logic App Standard** level, not on the workflow, and you can utilize the same **managed identity** across workflows, as shown in the following screenshot:

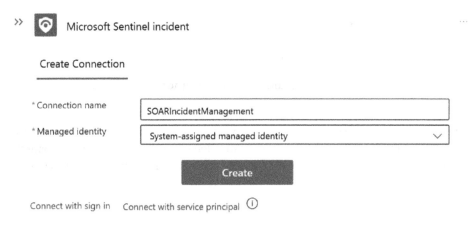

Figure 7.7 – Adding managed identity authentication

Once we have initialized our playbook and added a trigger, we can proceed with the next step.

Step B – adding the Entities - Get IPs action

In step B, we will add the **Entities - Get IPs** action:

Click on the plus (+) sign and select **Add an action**. Search for Microsoft Sentinel under **Azure**, and select **Entities - Get IPs** as the action. Add **Entities** from the dynamic content pane. The process is the same as with our hands-on examples in the previous chapter, as shown in the following screenshot:

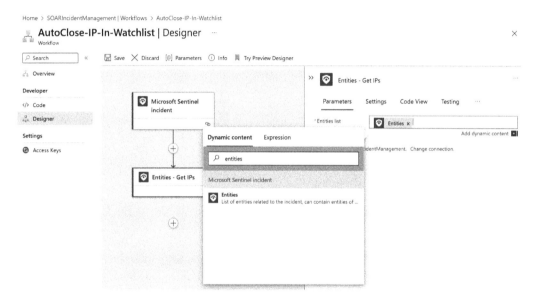

Figure 7.8 – Adding the Entities - Get IPs action in the workflow

Now that we have extracted IPs from the incident, we need to compare them with IPs in our watchlist.

Step C – getting allowed IPs from the watchlist

Step C involves querying our watchlist and checking whether the incident IP is part of our **AllowedIP** watchlist:

1. In the playbook, add a new action. Search for Azure Monitor Logs and select the **Run query and list results** action. We will get the information we need to authenticate the connection. The options are using a user identity or a service principal. In this case, we will use a service principal as shown in *Figure 7.9*:

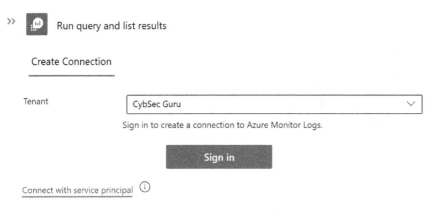

Figure 7.9 – Authenticating using a service principal

2. We will get an authentication window, where we need to enter service principal data, such as **Application ID**, **Tenant ID**, and **Secret**. To get those, we need to create a service principal. Open a new tab in the browser, and open the Azure AD admin portal (`aad.portal.azure.com`). From the left menu, open **Azure Active Directory** and then **App registrations**. Select **New registration** to start the wizard, as shown in the following screenshot:

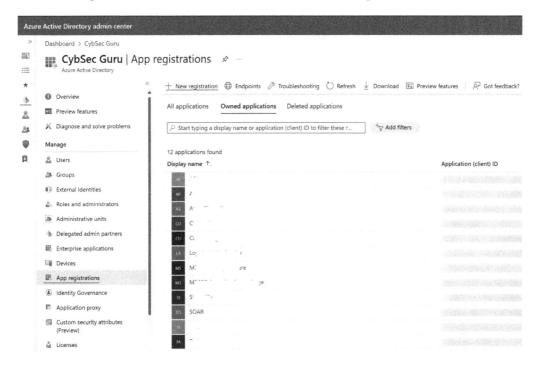

Figure 7.10 – Creating a service principal in Azure AD

3. Enter a name in the wizard (I used SOAR-LAReader) and select **Register**.

Once the application is registered, we need to copy **Application (client) ID** and **Directory (tenant) ID** to our playbook authentication:

I.   Copy **Application (client) ID** to **Client ID** in the playbook

II.   Copy **Directory (tenant) ID** to **Tenant**

III. Remember to add the connection name; I used the same name as for the service principal – SOAR-LAReader.

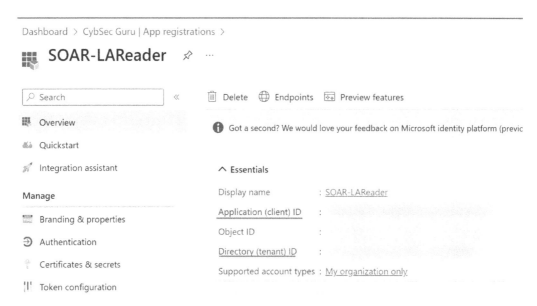

Figure 7.11 – The application and tenant IDs of the service principal

Figure 7.12 shows where to copy **Application (client) ID** and **Directory (tenant) ID** in the playbook.

>> 📊 Run query and list results

Create Connection

* Connection name	SOAR-LAReader
Client ID	2 d
Client Secret	Client secret of the Azure Active Directory application.
Tenant	4 5d

Create

Connect with sign in

Figure 7.12 – The application and tenant IDs in the workflow action

4. Now, we need to create and add a secret. Go back to the browser tab where we created the service principal and, from the menu, select **Certificates & secrets**. Select **New client secret** and add a description of the secret. I used the same name as for the connection name – SOAR-LAReader. Select **Add**, as shown in the following screenshot:

Figure 7.13 – Creating a secret for the service principal

5. Our secret will be added, and we need to copy the content of the **Value** field to the **Client secrets** field in our playbook, as shown in *Figure 7.14*:

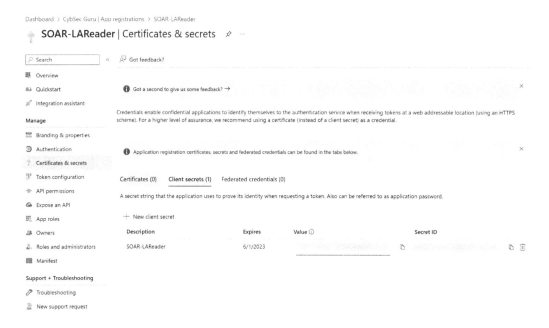

Figure 7.14 – Copying the secret value

Figure 7.15 shows where to copy to **Client Secret** in the playbook.

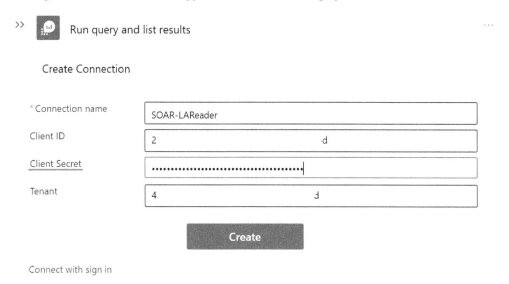

Figure 7.15 – Copying the secret value of the workflow action

6. Don't click on **Create** yet, as we will first assign permissions to be able to read logs. Open the new browser tab and open the Azure portal (`portal.azure.com`). Search for and select **Resource groups**. Select a resource group from the list; in my case, it is **SOAR**. Select **Access control (IAM)** from the menu and select **Add | Add role assignment**, as shown in the following screenshot:

Figure 7.16 – Adding a new role assignment

7. In the **Role** tab, search for and select **Log Analytics Reader**, as shown in *Figure 7.17*:

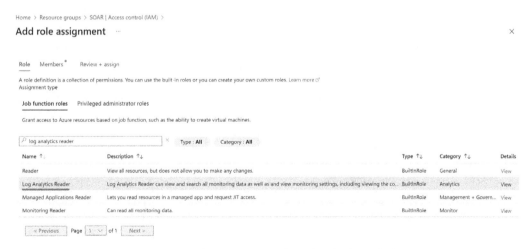

Figure 7.17 – Choosing Log Analytics Reader from the list

8. Click on **Next**. In the **Members** tab, click on **Select members** and search for your service principal – in my case, **SOAR-LAReader**. Click on **Select** and then on **Next**, as shown in the following screenshot:

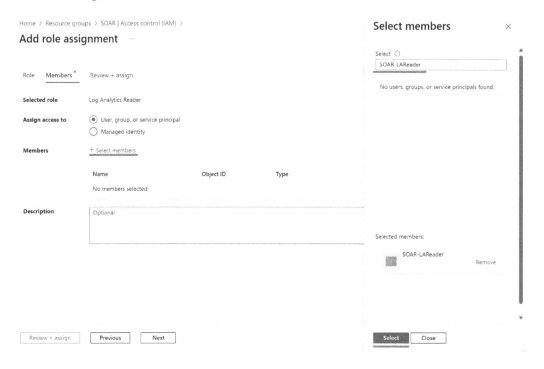

Figure 7.18 – Choosing the service principal

9. On the **Review + assign** tab, check the configuration and click the **Review + assign** button. The role will be assigned.

10. Now, go back to the playbook and select **Create**. We assigned permissions before selecting **Create** because now, from the drop-down menu, we can select a subscription, resource group, and resource name. Without permissions, the dropdown will be empty, and this data must be entered manually or dynamic values must be used.

11. After selecting a subscription, resource group, and resource name where Microsoft Sentinel is deployed, for **Resource Type**, select **Log Analytics Workspace**, and for **Time Range**, select **Last 24 hours**.

12. In **Query**, write the following:

```
_GetWatchlist('AllowedIP')
| where SearchKey == ""
```

On the second line, after ==, in the " " space, add **IPs Address** from the dynamic content. As in previous hands-on exercises, this will create a **For each** loop, as we can have more than one IP, as shown in the following screenshot:

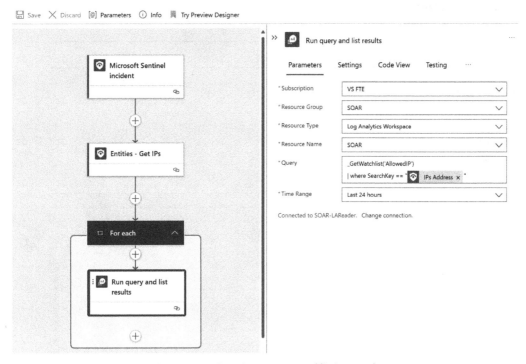

Figure 7.19 – Running a query and listing results

Note that if you used the **MaliciousIP** watchlist, you need to change the alias of the watchlist in the query.

Step D – comparing the IP from the incident with the watchlist values

In step D, we will compare the IP from the incident with the IP from the watchlist:

1. Select the plus sign (+) in the **For each** loop to add a new action, search for Control under **Built-in**, and select **Condition** under **Actions**, as shown in the following screenshot:

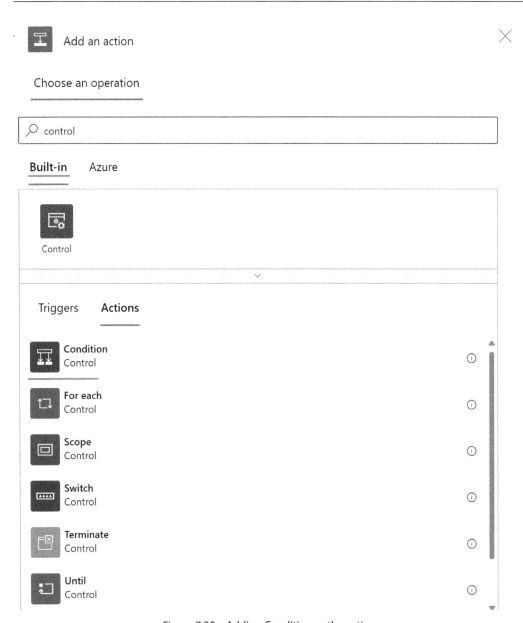

Figure 7.20 – Adding Condition as the action

2. Under **Condition**, fill in the details as follows:

I.   For the first value, select **Expression** (next to **Dynamic content**) and type length().
     Put the cursor inside the brackets of the length() expression, and then click on
     Dynamic content and select **value**. Select **OK**, as shown in *Figure 7.21*:

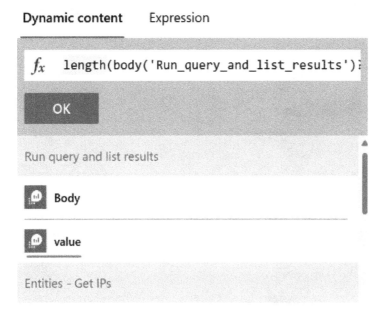

Figure 7.21 – Utilizing expressions with dynamic content

II.   For the second value, choose **is greater than** from the dropdown.

III. For the last value, add **0** (zero).

What this will do is check whether the watchlist query contains an IP (greater than zero) or
not. This is shown in *Figure 7.22*:

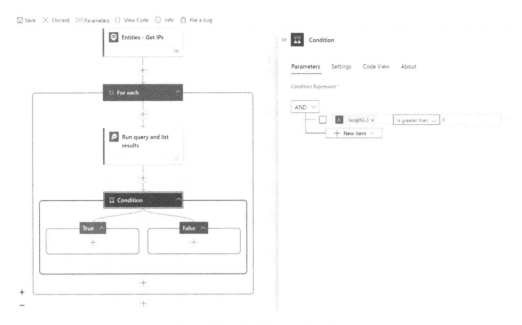

Figure 7.22 – Condition action values

Let's first configure the **True** stream of our condition.

Step E – configuring True stream

Next, we want to complete the **True** stream, where we want to auto-close the incident and add the `auto-close` tag:

1. Under **True**, select the plus sign (+), search for `Microsoft Sentinel`, and select the **Update incident** action.

2. For **Incident ARM id**, add **ARM ID** from **Dynamic content**.

3. In **Tags to add tag - 1**, add **auto-closed allowed IP**.

4. Change **Status** to **Closed**.

5. For **Classification reason**, you can put **BenignPositive - SuspiciousButExpected**.

6. For **Close reason text**, add **Auto-closed by AutoClose-IP-In-Watchlist playbook**.

7. **Classification reason** and **Close reason text** will appear only after you set **Status** as **Closed**.

Figure 7.23 shows the **Update incident** action with added values.

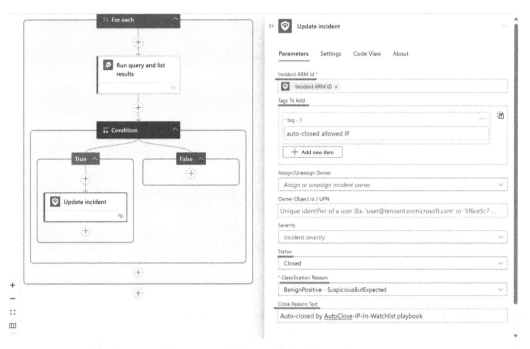

Figure 7.23 – Updating the incident values

After we have done the configuration of the **True** stream, we can proceed to the **False** stream of our condition.

Step F – configuring the False stream

In the **False** stream, we want to add a comment that an IP is not on the **AllowedIP** watchlist:

1. Under **False**, select the plus sign, search for `Microsoft Sentinel`, and select the **Add comment to incident** action.

2. For **Incident ARM ID**, add **ARM ID** from **Dynamic content**.

3. For the **Add comment to incident** message, add **IP address not found in the AllowedIP watchlist**. Then add **IPs Address** from **Dynamic content**, as there can be more than one IP.

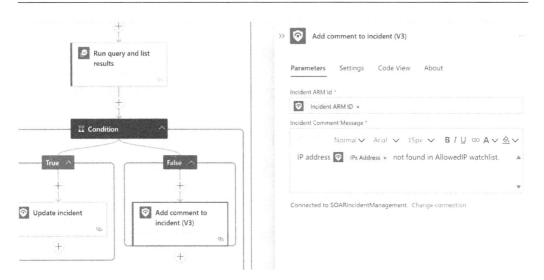

Figure 7.24 – Adding a comment to the incident values

Step G – assigning permissions to a managed identity

Our workflow is now created. Save it, and let's go to step G of the playbook creation process:

1. Go back to Logic App Standard, and from the left menu, select **Identity**, as shown in the following screenshot:

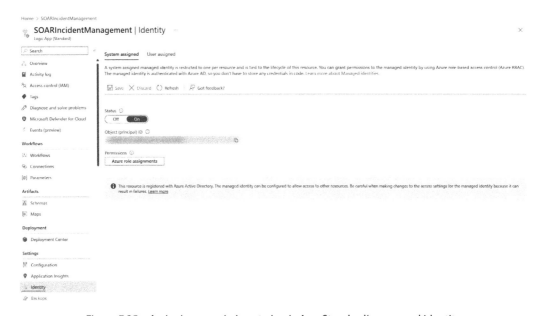

Figure 7.25 – Assigning permissions to Logic App Standard's managed identity

2. The rest of the process is the same as with the playbooks from the previous chapter. Select **Azure role assignments**, then **Add role assignment**, and add the **Microsoft Sentinel Responder** role.

As this is the last step in the playbook creation process, now we are ready to test our playbook.

Testing a playbook

The next six steps will guide you through the process of testing our playbook:

1. Go to the Microsoft Sentinel environment.

   We already have a detection rule created called **Test - Malicious IP**, and we can mark it and select **Enable** from the **Analytics** tab. Once it is enabled, wait for 15–20 seconds, and disable it so that it's not created every 5 minutes if not needed to test the automation rule.

2. Go to the **Incidents** tab, and there should be an incident called **Test - Malicious IP**.

   Right-click on the incident and select **Run playbook**.

   You will see the **Standard** playbook from the list of playbooks, which you can filter by plan. You will also see that the name looks different from the **Consumption** playbooks. The name will look like that of **Logic App Standard/workflow**. In my case, it is **SOARIncidentManagement/AutoClose-IP-In-Watchlist**. Select **Run**, as shown in the following screenshot:

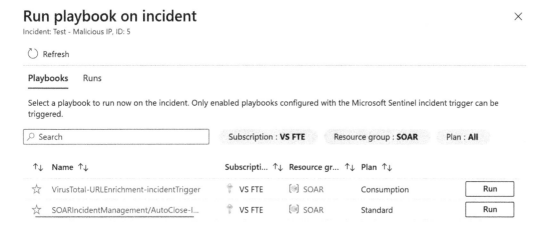

Figure 7.26 – Running the Logic App Standard workflow on the incident

3. If we refresh our **Incident** page, we will see that our incident has disappeared, as in the following screenshot:

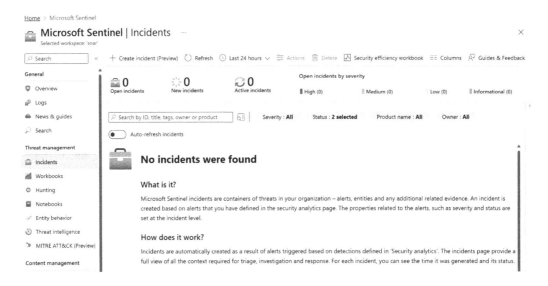

Figure 7.27 – The incident is closed using the playbook

4. Filter the **Incident** page by status and include **Closed**. Our incident will reappear. Select it, and from the details on the right-hand side of the page, we can see our tag and that the status is **Closed**, as shown in the following screenshot:

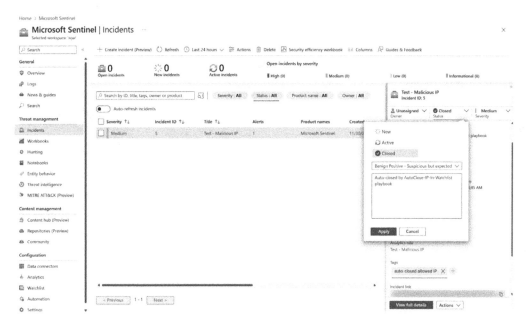

Figure 7.28 – Values changed using the playbook

5. To test whether an IP is on the watchlist, go and change the IP value on the watchlist, as shown in the following screenshot:

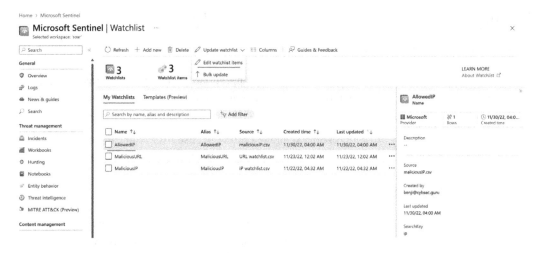

Figure 7.29 – Editing our watchlist values

6. We can change the IP from 45.81.226.17 to 45.81.226.181 and save it. Wait a few minutes until some info is seen in **Logs**, recreate the incident, and rerun the playbook. In this case, it will add a comment to the incident, as shown in the following screenshot:

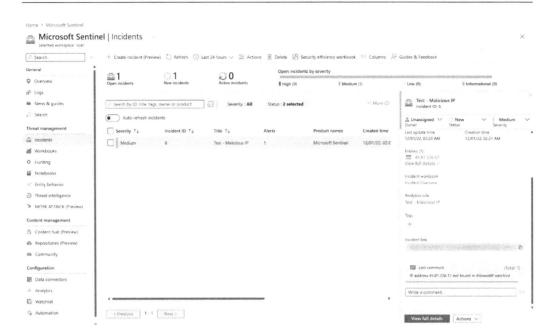

Figure 7.30 – The comment value on the playbook's second run

Now that we have successfully tested our playbook, we can continue to the next hands-on example.

Closing an incident based on SOC analyst input

Before we begin, you will need the following:

- You need to have access to Microsoft Sentinel with appropriate permissions (Microsoft Sentinel Contributor, Logic App Contributor, and permission to assign RBAC controls – Owner or User Access Administrator).

- You will need at least one Microsoft Exchange Online license user. You can add a trial for Microsoft Office 365 using the same tenant of your Azure subscription.

Creating a playbook

In the previous example, we saw how we can create a playbook to auto-close incidents so that SOC analysts don't lose time investigating an IP address at all. While this is great when we have one IP address in an incident, it can be problematic if there are multiple IPs and some are on the watchlist while some are not. In our previous example, the incident would auto-close even if there were multiple detected IPs in an incident, but only one is in our watchlist. So, if there are five IP addresses in the incident, four of which are not on the **AllowedIP** watchlist and one is on the watchlist, the playbook will still auto-close the incident, but it will add comments to the incident saying that IP addresses are not on the watchlist.

In this example, we will adjust our playbook to send an approval email to a SOC analyst with a list of approved and unapproved IPs.

We will do the following:

1. Make sure that our **AllowedIP** watchlist has `45.81.226.17` as the IP value.

2. Edit our `MaliciousIP` watchlist and add more IPs that will be used in the detection rule.

3. Create a playbook:

    I. **Step A**: We will run the playbook on the incident, so we will need to utilize the incident trigger.

    II. **Step B**: Then, we need to initialize two variables that will contain allowed and not-allowed IPs.

    III. **Step C**: We need to extract IPs from the **Entities** array.

    IV. **Step D**: Get IPs from the watchlist using **Azure Monitor Logs** and compare them to IPs from incidents. For this, we will use the service principal connection from the previous exercise.

    V. **Step E**: Compare an IP from the watchlist with the IP from the incident.

    VI. **Step F**: If the IP is in the **AllowedIP** watchlist, add it to the allowed variable.

    VII. **Step G**: Otherwise, add it to not-allowed variable.

    VIII. **Step H**: Send an approval email to the SOC analyst with a list of all IPs and whether they are approved or not.

    IX. **Step I**: Based on the user's decision, close the incident with a tag; otherwise, add a comment, assign the incident to the user who responded to the email, and mark the incident as **Active**.

    X. **Step J**: We will use the managed identity of Logic App Standard and permissions already applied.

Let's start.

As the first step, if you have changed an IP in the **AllowedIP** watchlist, return to the starting value – `45.81.226.17`.

As the second step, in the **MaliciousIP** watchlist, let's add four more IPs:

1. Go to **Watchlist**, select the **MaliciousIP** watchlist, and select **Update watchlist | Edit watchlist items**.

2. Click on **Add new** and add the `45.81.226.22` value. Do the same with three more IPs:

    - `45.81.226.55`
    - `45.81.226.88`
    - `45.81.226.99`

3. Click on **Save** as shown in the following screenshot:

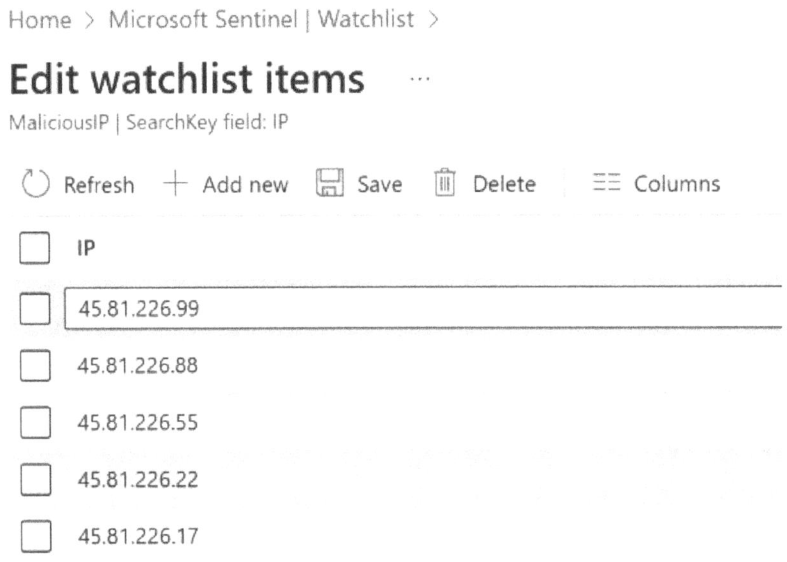

Home > Microsoft Sentinel | Watchlist >

Edit watchlist items   ...

MaliciousIP | SearchKey field: IP

⟳ Refresh   + Add new   💾 Save   🗑 Delete   ☰☰ Columns

☐ IP

☐ 45.81.226.99

☐ 45.81.226.88

☐ 45.81.226.55

☐ 45.81.226.22

☐ 45.81.226.17

Figure 7.31 – The Edit watchlist items values

4. After a few minutes, check **Logs** to ensure all IPs are written to our watchlist.

5. Once all five IPs are in **Logs**, let's go and rerun our **Test - Malicious IP** analytic rule (use the **Enable** rule and disable after 15–20 seconds).

6. If we go to the **Incidents** tab, we will see our new incident with five IPs in it now.

The third step is to create our playbook!

Step A – initializing a playbook and adding trigger

In step A, we will create a new workflow and add an incident trigger to it:

1. Let's go to the **Logic App Standard** resource we created in the previous exercises. On the left menu, select **Workflows**, then **Add**, enter a workflow name (**Approval-AutoClose-IP-In-Watchlist** in my case), select the **Stateful** state type, and select **Create**.

2. Our new workflow will be added. Select the new workflow to start building it.

    As a trigger, we will continue to use the **Microsoft Sentinel** trigger. It will automatically start utilizing the managed identity we created in the previous exercise – **SOARIncidentManagement** – and it has the **Microsoft Sentinel Responder** role already assigned.

Step B – initializing variables

In step B, we will initialize two variables:

1. Select the plus sign (+) to add a new action, search for Variables under **Built-in**, and select the **Initialize variable** action. The name will be **AllowedIP** and the type will be **String**, as shown in the following screenshot:

>>   {x}   Initialize variable   . . .

Parameters	Settings	Code View	About

* Name

AllowedIP

* Type

String   ∨

Value

Enter initial value

Figure 7.32 – Initializing the variable values

2. Add one more, but now enter **NotAllowedIP** in the **Name** field.

    One thing to note is that the names of the variables in the list are **Initialize variable** and **Initialize variable 2**. When we need to change something, we must open both actions to examine each one in more detail. *Figure 7.33* shows initialized variables.

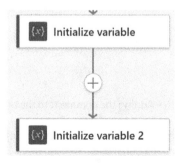

Figure 7.33 – Automatic naming in Logic Apps

3. To make this better, we can rename each action, and I always try to assign a name based on what the action is about. So, for the first, it would be **Initialize variable- AllowedIP**, and for the second, **Initialize variable- NotAllowedIP**. To rename the action, select it, select the **Name** field at the top (**Initialize variable**), and add text, as shown in the following screenshot:

Figure 7.34 – Renaming the action

Alternatively, we can select the three dots and **Add a comment**, where we can write what the action is about, as shown in the following screenshot:

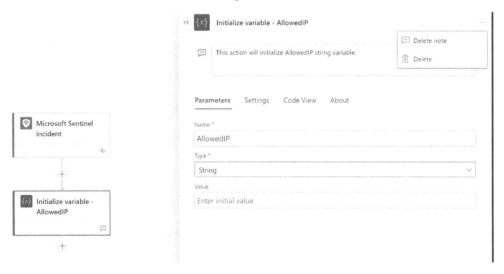

Figure 7.35 – Adding the comment to the action

If one or two words can describe the action, I prefer renaming. Comment only when there is a need to describe more.

We will use the same approach to rename the second variable to **Initialize variable- NotAllowedIP**.

Step C – adding the Entities – Get IPs action

Now, we will add the Microsoft Sentinel **Entities - Get IPs** action to get IPs from incidents by adding **Entities** from **Dynamic content**.

Step D – getting IPs from the watchlist

In step D, we will get IPs from the watchlist we created:

1. We will add the **Azure Monitor Logs Run query and list results** action with the same details as in the previous hands-on exercise in this chapter. The query we used was as follows:

```
_GetWatchlist('AllowedIP')
| where SearchKey == ""
```

2. In the second line, after == in the " " space, add IPs Address from the dynamic content. As in previous exercises, it will create a For each loop, as we can have more than one IP.

Step E – comparing the IP from the incident with the watchlist values

In step E, we will compare IP from the incident with the IP from the watchlist:

1. We will use the same condition by typing length() into the expression and then adding value from the dynamic content.

2. The second value will be **is greater than**, and the third value will be **0** (zero), as shown in the following screenshot:

Figure 7.36 – Condition action values

Let's configure the **True** stream of our condition.

Step F – configuring the True stream

Now comes a different approach. In **True**, select the plus sign to add a new action, and from **Actions**, choose **Variables - Append to a string variable**. What this action does is ensure that each IP that is on the **AllowedIP** watchlist will be added to this variable:

- For the name, we will select **AllowedIP** from the list, and for **Value**, we will select **IPs Address** from **Dynamic content** and add a comma and empty space after it (,   ). This is to ensure separation when we have multiple values. We will also rename it and add **AllowedIP** in the name as shown in *Figure 7.37*:

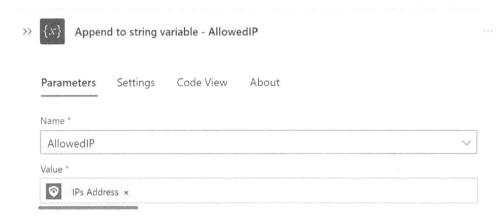

Figure 7.37 – Appending to the variable values

After the **True** stream is configured, let's proceed to the **False** stream.

Step G – configuring the False stream

In **False**, select the plus sign to add a new action, and from **Actions**, choose **Variables - Append to string variable**. What this action does is add each IP that is not in the **AllowedIP** watchlist to this variable:

1. For name, we will select **NotAllowedIP** from the list.

2. For **Value**, we will select **IPs Address** from **Dynamic content** and add a comma and empty space after it (,   ).

3. We will also rename it and add **NotAllowedIP** to the name. The full name will be **Append to string variable - NotAllowedIP**.

Figure 7.38 shows the **True** and **False** streams:

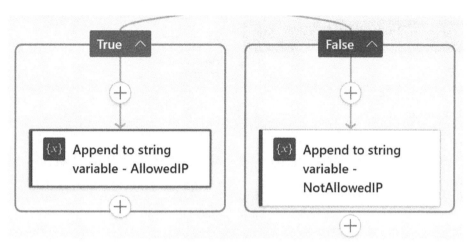

Figure 7.38 – True and False condition streams

Let's proceed now to step H and add the approval process.

Step H – sending an approval email

So, we have now added our IPs in the appropriate variables, and we need to send an approval email to the SOC analyst:

1. Outside of the **For each** loop, click on the plus sign to add a new action.
2. Search for the Office 365 Outlook connector, and then for the Send approval email action. We will need to create a connection using a user identity, and that user must have a Microsoft Exchange Online license. If you don't have one, use an Outlook.com account and an Outlook.com connector.

3.   Select **Sign in** and sign in as your user, as shown in the following screenshot:

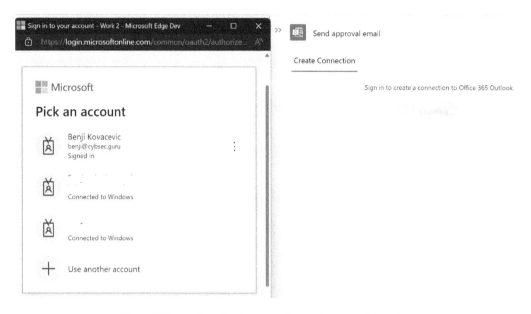

Figure 7.39 – Authenticating an action using a user's identity

4.   The first thing that we need to do is to select **Add new parameter** and then select **Body**.

Then, we need to fill in our fields:

- **To**: Enter the email address that you will send the approval email to. Normally, this will be the incident owner or dedicated SOC mailbox.

- **Subject**: **Approval Request for incident** (and add **Incident Title** from **Dynamic content**).

- **User Options**: **Auto-close**, **Further investigation needed**.

- **Body**: The body can include the following:

  - **New incident detected!**

  - **Incident title**: Add an incident title from **Dynamic content**

  - **Incident severity**: Add an incident severity from **Dynamic content**

  - **Incident description**: Add an incident description from **Dynamic content**

  - **Incident URL**: Add an incident URL from **Dynamic content**

  - **List of IPs in AllowedIP watchlist**: Select the **AllowedIP** variable from **Dynamic content**

  - **List of IPs not in AllowedIP watchlist**: Select the **NotAllowedIP** variable from **Dynamic content**

- **Importance**: High

- **Hide HTML message**: **Yes**. This is needed so that we get the **User Principal Name (UPN)** of the user that responded to the approval process

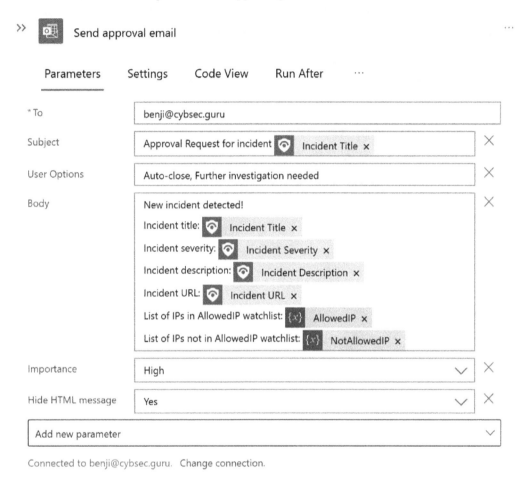

Figure 7.40 – The Send approval email values

Once we have configured the approval process, we need to configure what will happen once one of the selections is made in the approval.

Step I – SOC analyst decision evaluation

The next step is to take action based on the SOC analyst's input. If the SOC analyst chooses **Auto-close**, we will close the incident. If the SOC analyst chooses **Further investigation needed**, we will assign the incident to that user and add a comment, with information about a list of IPs on the **AllowedIP** watchlist and a list of those that are not. We will use the **Condition** action here again:

1. Let's click on the plus sign (+) below the **Send approval email** action and then click on **Add new action**. Search for Control again and choose **Condition**:

   - Rename it **Condition - analysts input**
   - For the first value, choose **SelectedOption** from the dynamic content
   - The second value will stay as **is equal to**
   - The third value will be **Auto-close**

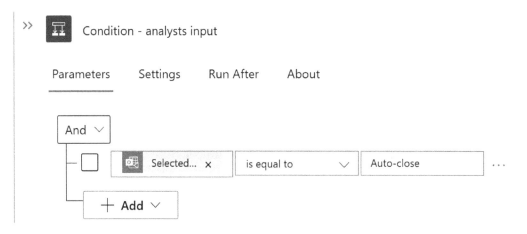

Figure 7.41 – Condition actions based on the analyst's input

2. In **True**, we will add a new action. Search for **Microsoft Sentinel** and select **Update incident** action:

   I.   Rename the action **Update incident – Auto-close**.

   II.   Add **Incident ARM ID** from the dynamic content.

   III. Change **Status** to **Closed**.

   IV.   **Classification reason** can be **BenignPositive – SuspiciousButExpected**.

   V.   **Close reason**: **Auto-closed using the playbook**. **Auto-closed by** (here, we add UserEmailAddress using **Dynamic content** – under the **Send approval email** action details).

Please note that to get the email addresses of the user that responded, we have to hide the HTML message in the **Send approval email** action:

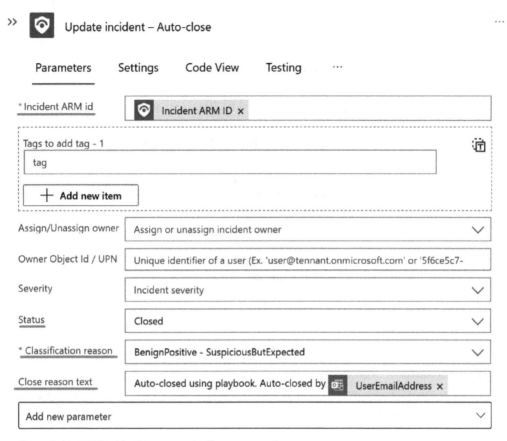

Figure 7.42 – Updating the incident values when auto-closing

If we want, we can also add a comment action and also comments with a list of IPs that are in the watchlist, as well as those that are not.

3. In **False**, we will add a new action by going to **Microsoft Sentinel | Add comment to incident**, as shown in *Figure 7.43*:

    I.   Add **Incident ARM ID**, and then in the body, add the following:

- **List of IPs in AllowedIP watchlist**: Select the **AllowedIP** variable from the dynamic content

- **List of IPs not in AllowedIP watchlist**: Select the **NotAllowedIP** variable from the dynamic content

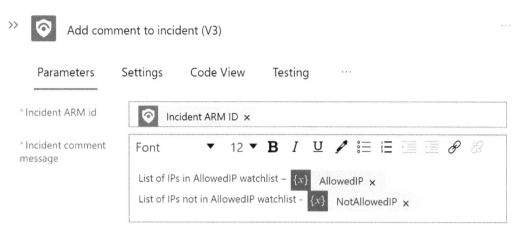

Figure 7.43 – Adding a comment to the incident values

II. Add a new action after **Add comment to incident** by going to **Microsoft Sentinel | Update incident**.

III. Rename the action **Update incident – assign the incident to the responder**.

IV. Add **Incident ARM ID**.

V.   Change the **Assign/Unassign owner** field to **Assign**, and in **Owner Object Id/UPN**, add **UserEmailAddress** using the dynamic content – under **Send approval email action**, as shown in the following screenshot:

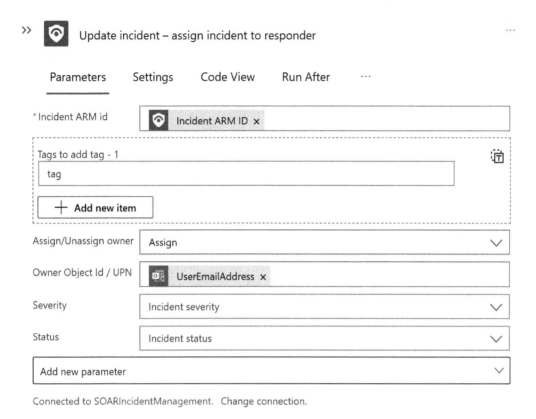

Figure 7.44 – Updating the incident values to assign an owner

Now, we can save our playbook and test it out!

Testing a playbook

The last step in our process is to test the playbook and make sure it runs with no errors. Following are the steps to do that:

1. To test the playbook, we will go to **Microsoft Sentinel** and the **Incident** tab. We will have the **Test - Malicious IP** incident. Right-click on it and select **Run playbook**.

2. From the list, search for and run the playbook (**SOARIncidenManagement/ Approval-AutoClose-IP-In-Watchlist**).

3. Go to **Outlook Web** (outlook.office.com/mail) and sign in as the user to whom we are sending the approval request (the **TO** field in the **Send approval email action**). We will see an email in the inbox from the user who authenticated the **Office 365 Outlook** connector, with the subject **Approval Request for incident Test - Malicious IP**, as shown in the following screenshot:

Figure 7.45 – An approval email example

We can see that the text is not formatted and is hard to read. In *Chapter 9*, we will go through tips and tricks when creating a playbook, and one of the tips will be how to utilize the **Compose** action and basic HTML to make it user-friendly.

4. Let's select **Auto-close** this time. We will get a new box to confirm **Auto-close**, and we will select it. There will be a notification that the action is completed. Refresh the view if **Auto-refresh incidents** is not on, and we will see that our incident is not on the list anymore.

5. Let's filter it by closed incidents, select our incident, and check the summary on the right. We can see that it is closed and our closed comment, as shown in the following screenshot:

Figure 7.46 – An incident update if the analyst selects Auto-close

6. Let's now test what will happen if we choose **Further investigation needed**. We can reopen our incident by changing the status to **New** or **Active**, or recreate the incident by enabling and disabling the analytic rule, as shown in the following screenshot:

Figure 7.47 – An incident update if the analyst selects Further investigation needed

Usually, after testing, we would assign this kind of playbook to the automation rule so that it is run upon incident creation. We want to make this process as automatic as it can be.

Let's now see an example of how we can automatically close incidents using automation rules without a playbook or user interaction.

Auto-closing incidents using automation rules

Before we begin, you will need to have access to Microsoft Sentinel with appropriate permissions (Microsoft Sentinel Contributor).

Creating an automation rule

In this case, we want to close an incident using the automation rule on incident creation, when a specific IP is detected during a specific time window. One such scenario is if we are doing penetration testing, and this IP address will create a lot of false positives. Therefore, we don't want to overload SOC analysts with these incidents and instead let them focus on their day-to-day operations. The following case will involve penetration testing with the SOC not engaged:

1. To begin, we need to go to Microsoft Sentinel and the **Automation** tab.

2. Select **Create** and choose **Automation rule**.

3. Under **Automation rule name**, we can add **Pen-Testing False Positive**.

4. The trigger should stay as **When incident is created**.

5. In **Conditions**, we can leave **All** for **Analytic rule name** or select a specific one. We can select only one. We will be using **Test - Malicious IP**.

6. We will click on **Add** and select **Condition (Add)**:

   - For the first value, select **IP address**.

   - The second value will be **Contains**.

   - For the third value, enter our IP address – 45.81.226.17.

   Figure 7.48 shows conditions configured in steps 5 and 6.

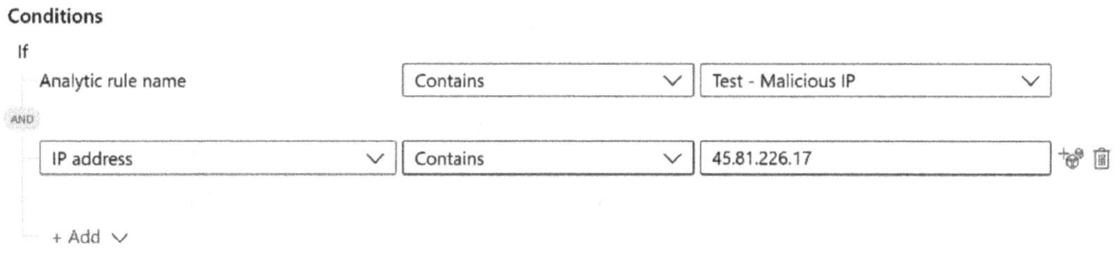

Conditions

If

Analytic rule name	Contains ∨	Test - Malicious IP ∨

AND

IP address ∨	Contains ∨	45.81.226.17

+ Add ∨

Figure 7.48 – The automation rule condition values

7. We will not add any new conditions, but you can test utilizing many other filters.

   In **Actions**, we will select **Change status**, choose **Closed**, and then **Classification (Benign Positive - Suspicious but expected)**. Add a comment, **Pen-testing incident**, as shown in the following screenshot:

Actions ⓘ

Change status	∨	🗑

✔ Closed	∨

Benign Positive - Suspicious but expected	∨

Pen-testing incident

Figure 7.49 – The automation rule action values

8. In **Rule expiration**, choose the date you are creating the automation rule, and for the time, enter the time that is 10–15 minutes in the future, as shown in the following screenshot:

Rule expiration ⓘ

| 12/03/2022 🗓 | 3:15 AM |

Order ⓘ

| 1 |

Figure 7.50 – The automation rule expiration values

9. Select **Apply**.

Let's test our automation rule – we have a 10–15-minute time window based on our configuration in **Rule expiration**.

Testing an automation rule

The following steps will guide you through the testing of the automation rule we created:

1. The next few steps will guide you through the testing of the automation rule we created. We will utilize our **Test - Malicious IP** analytic rule (by going to **Microsoft Sentinel | Analytics**) and enable and disable the rule so that an incident gets generated. Let's go to the **Incident** tab and select **Closed** as the status. We will see our incident and that it is automatically closed, as shown in the following screenshot:

Figure 7.51 – An example of the incident auto-closed with the automation rule

2. Let's wait until our configured rule expiration time expires, open our automation rule, and then we will be able to see at the end that it is in **Disabled** mode now.

3. We can recreate our incident again, and we will see that it will not be closed.

4. To see who closed the incident, you can utilize the **Logs** tab and examine the following query:

```
SecurityIncident
| where IncidentNumber == "< enter your incident ID >"
| summarize arg_max(TimeGenerated, Status, Classification,
ClassificationComment, ModifiedBy)
```

Figure 7.52 shows a sample query and the results from step 4.

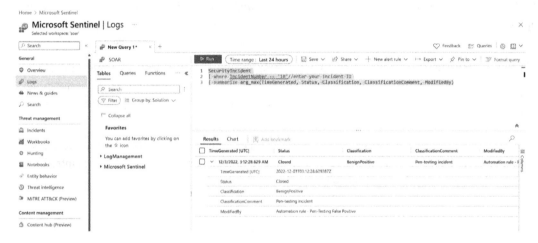

Figure 7.52 – Seeing raw data in the Logs tab

Here is the **incident ID (IncidentNumber** in KQL):

Figure 7.53 – Locating the incident ID in the Incidents tab

With automation rule creation and testing, we have finished our hands-on examples for *Chapter 7*.

Summary

In this chapter, we continued with hands-on examples and focused on how to manage incidents using automation.

In the first example, we used a playbook to close an incident when it contained a specific IP. This can be replicated with any piece of information that the incident contains, and we can use the incident title, the account, the host associated with the incident, and so on.

With the second example, we used the logic of example number one and expanded it with SOC analyst approval to auto-close the incident. This can be really important when we have multiple IPs for the same incident, as we don't want to auto-close if some IPs are not on the allowed list.

The final example in this chapter used automation rules to auto-close incidents for a specific analytic rule with specific IPs, during a specific time range. This can be helpful when we are performing penetration testing and we don't want to overload our SOC analysts, or if we are performing specific testing and those incidents will be false positives.

In the next chapter, we will perform the final hands-on examples, and they will focus on response automation. We will go over the importance of having the option to respond fast, and we will introduce some common use cases in our hands-on examples.

8

Responding to Incidents Using Automation

In the previous chapter, we focused on incident management using automation.

The first hands-on example was to auto-close an incident with no analyst interaction. We utilized the watchlist feature in Microsoft Sentinel, where we stored our allowed IP address and compared it with IPs involved in the incident. Based on the result, we auto-closed the incident or left a comment stating that the IP was not on the watchlist.

The second example expanded on the first example. As incidents can have more than one IP, we utilized an approval email action to ask analysts whether the incident should be auto-closed or whether a further investigation would be needed.

The final example used the automation rule to auto-close incidents on incident creation if an IP matches our specific IP. One example of using automation could be to auto-close incidents during penetration testing when the SOC is not a part of it.

This chapter will focus on how to respond to incidents utilizing automation. We usually want to utilize this to remove the middleman when responding to an incident, especially when we need to block an IP or user, or if we need to isolate the host.

This chapter will cover the following topics:

- Automating responses to incidents
- Blocking a user upon suspicious sign-in
- Isolating a machine upon new malware detection

Automating responses to incidents

In organizations, we have multiple teams handling different segments of IT. We have networking departments, system administration, device management teams, and so on. Traditionally, when we have an incident and the SOC needs to isolate a machine or block an IP in the firewall, they would need to raise a ticket in ITSM or send an email request. The department in question then picks this up and acts on it. However, this takes time. An analyst needs to log in to ITSM, open a new ticket, enter all data and a justification in detail, and submit it. Then, the appropriate department needs to verify and act on it, which means they will need to copy the IP or host details, go to the appropriate system, and update it. This adds many unnecessary steps.

With automation, this process can be more efficient. After SOC analysts do their investigation and realize that they need to isolate a machine or block an IP address, they can run a playbook that will perform that step for them. If an organization has a policy that needs approval from the appropriate department, it can add an approval process to the playbook. This will also automate networking or device management departments' processes, as they need to approve the process, and the playbook will then perform this process for them. There's no need for anyone to copy data and double-check that the correct data is entered. All data will be automatically extracted from the incident and utilized in the playbook.

When we deal with high-severity incidents, every second counts, especially nowadays when those seconds can mean you stopping the spread of incidents such as ransomware. This is a reality thanks to automation, and organizations are starting to utilize it for incident responses regularly.

So, should we utilize automation for incident responses? Absolutely! Let's look at our examples and see how we can do it!

Blocking a user upon suspicious sign-in

In this example, we will block users upon an unknown sign-in detected by our system. This can be for different reasons, but in our example, we will have a use case involving a user signed in from an IP address from our **MaliciousIP** watchlist.

Before we begin, you will need the following:

- You need to have access to Microsoft Sentinel with appropriate permissions (Microsoft Sentinel Contributor, Logic App Contributor, and permission to assign RBAC controls – Owner or User Access Administrator)

- Global Administrator or User Administrator role permissions in Azure **Active Directory** (**AD**)

Creating a playbook

Let's see how to create a playbook:

1. Enable the Azure AD data connector in Microsoft Sentinel.

2. We need an additional user, who we will block.

3. We need detection to create an incident on which we will run our playbook.

4. We need a playbook with the following attributes:

   - Step A – create a playbook with an incident trigger

   - Step B – get user and IP entities from the incident

   - Step C – add an Azure Monitor Logs action

   - Step D – compare an IP from the watchlist with the IP from the incident

   - Step E – block the user if the IP is in the watchlist and leave a comment on the incident

   - Step F – assign permissions to a managed identity

Let's start with step 1.

Step 1 – enable the Azure AD connector

First, we will enable the Azure AD data connector to send sign-in data to Microsoft Sentinel:

1. Go to **Microsoft Sentinel | Data connectors**, search for and select **Azure Active Directory**, and then select **Open connector page**, as shown in the following figure:

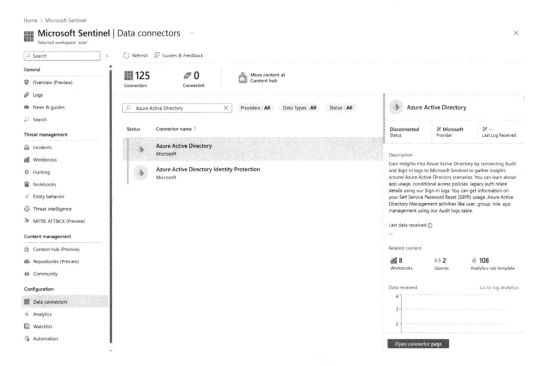

Figure 8.1 – The Data connectors tab

2.   Under **Configuration**, select **Sign-In Logs** and then **Apply Changes**, as shown in *Figure 8.2*:

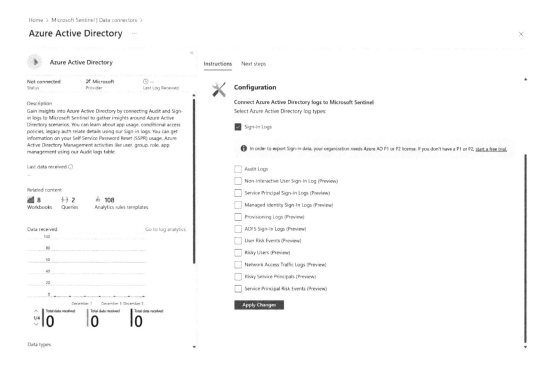

Figure 8.2 – Connecting to Sign-In Logs in the Azure AD connector

Now we have enabled the Azure AD data connector, and we will get all users' sign-in logs from the connected tenant.

Step 2 – create a test user

Step 2 is to create an additional user that we will utilize to test our playbook on:

1.   Go to the Azure AD portal (`aad.portal.azure.com`) and select **Azure Active Directory | Users | New user | Create new user**, as shown in the following screenshot:

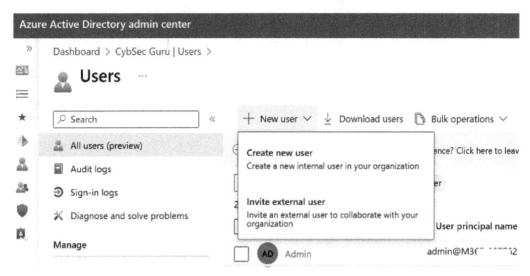

Figure 8.3 – Creating a new user

2. Enter a username, name, and password, as shown in the following screenshot:

Figure 8.4 – New user details

3.   Let's sign in with this user to verify their creation and generate one sign-in event for our data connector. On the **InPrivate browser** tab, go to `portal.office.com` and sign in as our test user.

4.   After a few minutes, go to **Microsoft Sentinel** | **Logs** and search for our latest sign-in in the `SigninLogs` table, as shown in the following screenshot:

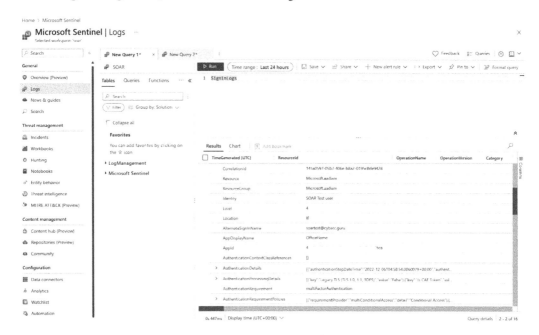

Figure 8.5 – User sign-in in Microsoft Sentinel Logs

Now we have a connector enabled, a user created, and one successful sign-in as the log in our `SigninLogs` table.

Step 3 – create a detection rule

Step three is to create a detection rule:

1.   Go to **Microsoft Sentinel** | **Watchlist**, select the **MaliciousIP** watchlist, and then go to **Update watchlist** | **Edit watchlist items**. Make sure that the only IP in the watchlist is `45.81.226.17`. If there are any others, remove them.

2. Next, go to **Microsoft Sentinel | Analytics** and select **Create | Scheduled query rule**. In the **General** tab, add `Test - Suspicious sign-in` as a name and mark **Status** as **Disabled**. In the **Set rule logic** tab, for **Rule query**, add the following:

```
_GetWatchlist('MaliciousIP')
| extend UsrAccount = "SOARTest"
| extend UPNSuffix = "cybsec.guru"
```

Here, `UsrAccount` is the username of the user, and `UPNSuffix` is your domain.

3. In **Entity mapping**, select **IP** for **Entity type**, **Address** for **Identifier**, and **SearchKey** for **Value**. Then, click on **Add entity**, and select **Account** for **Entity type**, **Name** for **Identifier**, and **UsrAccount** for **Value**. Finally, click on **Add identifier**, and select **Account** for **Entity type**, **UPNSuffix** for **Identifier**, and **UPNSuffix** for **Value**, as shown in the following screenshot:

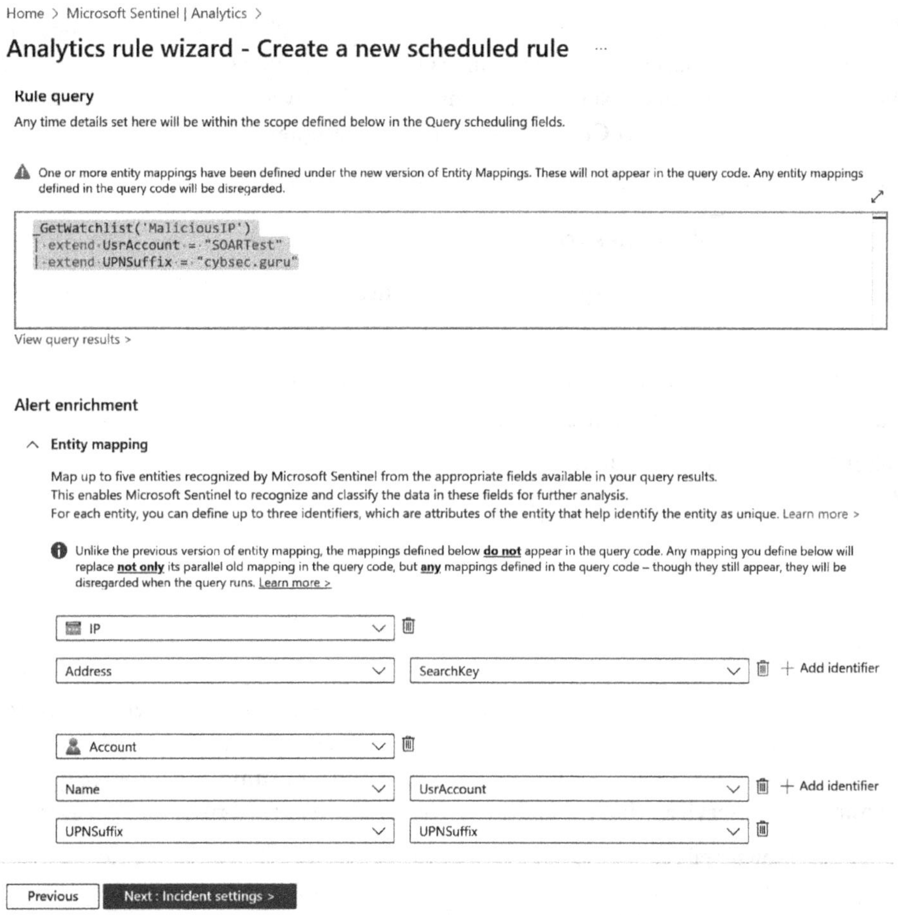

Home > Microsoft Sentinel | Analytics >

Analytics rule wizard - Create a new scheduled rule   ⋯

Rule query

Any time details set here will be within the scope defined below in the Query scheduling fields.

⚠ One or more entity mappings have been defined under the new version of Entity Mappings. These will not appear in the query code. Any entity mappings defined in the query code will be disregarded.

```
_GetWatchlist('MaliciousIP')
| extend UsrAccount = "SOARTest"
| extend UPNSuffix = "cybsec.guru"
```

View query results >

Alert enrichment

⌃ **Entity mapping**

Map up to five entities recognized by Microsoft Sentinel from the appropriate fields available in your query results.
This enables Microsoft Sentinel to recognize and classify the data in these fields for further analysis.
For each entity, you can define up to three identifiers, which are attributes of the entity that help identify the entity as unique. Learn more >

ⓘ Unlike the previous version of entity mapping, the mappings defined below **do not** appear in the query code. Any mapping you define below will replace **not only** its parallel old mapping in the query code, but **any** mappings defined in the query code – though they still appear, they will be disregarded when the query runs. Learn more >

🖥 IP	⌄	🗑		
Address	⌄	SearchKey	⌄ 🗑	+ Add identifier

👤 Account	⌄	🗑		
Name	⌄	UsrAccount	⌄ 🗑	+ Add identifier
UPNSuffix	⌄	UPNSuffix	⌄ 🗑	

Previous | Next : Incident settings > |

Figure 8.6 – Creating a new analytics rule

4. In **Query scheduling**, change the **Run query every** setting from **Hours** to **Minutes**.

5. Proceed to **Review and create**, verify the configuration, and select **Create**. If you want to test your playbook while adding each step, enable and disable the rule you created so that you can run the playbook on it.

Step three is done. Let's proceed to step four and create our playbook.

Step 4 – create a playbook

Now that we have configured our test user and detection rule, we can create a playbook.

Step A – initialize a playbook with an incident trigger

First, we need to initialize a playbook, and in this case, it will be with an incident trigger:

1. Go to **Microsoft Sentinel** | **Automation** and select **Create** | **Playbook with incident trigger**. Choose the correct subscription and resource group, and for **Playbook name**, enter `Block-AzureAD-User`. Click on **Next**, leave the managed identity as is, click on **Next** again, verify the configuration, and select **Create and continue to designer**.

    Our trigger is already initialized – **Microsoft Sentinel incident**.

Step B – get user and IP entities from the incident

After we have our trigger, we need to get user and IP entities from the incident, and for this, we will need to perform two actions:

1. Select **New step**, search for and select **Microsoft Sentinel**, and then select **Entities – Get IPs**. Add **Entities** from the dynamic content.

2. Select **New step** again, search for `Microsoft Sentinel`, and then select **Entities – Get Accounts**. Add **Entities** from the dynamic content.

Step C – add an Azure Monitor Logs action

Now, we need to check whether an IP is on our watchlist. Like in the examples in *Chapter 7*, we will use **Azure Monitor Logs**:

1. Select **New step**, search for **Azure Monitor Logs**, and then select the **Run query and list results** action. The service principal we created is connected to the **Logic App Standard** resource type, and we will need to create a new one for the **Logic App Consumption** resource type. Select **Connect with a service principal** and enter `SOARLAReader-Consumption` as the name.

2. In the new browser tab, go to `aad.portal.azure.com` and click on **Azure Active Directory** | **App registration**, and select the app we created in *Chapter 7* (SOAR-LAReader, in my case). Copy **Application (client) ID** to **Client ID** and **Directory (tenant) ID** to **Tenant** in the playbook.

3. From our app registration, go to **Certificates & secrets**, and you will see that the secret value we generated earlier is hidden and cannot be reused. Let's create a new secret, copy its value, and paste it into the **Client Secret** space in our playbook. You can save the secret value safely if you want to reuse it if needed – for example, by utilizing Azure Key Vault. Please note that if you delete a secret in the app registration, all playbooks or other services using app registration with that secret will not be able to perform the task defined in the playbook.

4. Select **Create**. We already applied **Log Analytics Reader** as a permission to select the same values as before. The change in the query is that we will now use the `MaliciousIP` watchlist:

```
_GetWatchlist('MaliciousIP')
| where SearchKey == ""
```

In the second line, after `==` in the `" "` space, add **IPs Address** from the dynamic content. As in previous exercises, this will create a **For each** loop, as we can have more than one IP in the incident, as shown in the following screenshot:

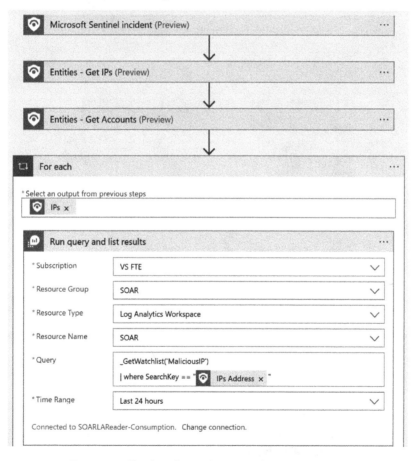

Figure 8.7 – Checking for an IP in our MaliciousIP watchlist

With this, we will get information about whether the IP in the incident on our list of malicious IPs.

Step D – add a Condition action

To make a true/false condition based on the results from step C, we will use a **Condition** action:

1. Select **Add an action** in the **For each** loop, search for `Control` under **Built-in**, and then select **Condition** from **Actions**.

2. Under **Condition**, fill in the values as follows:

   I. For the first value, select **Expression** (next to **Dynamic content**), and type `length()`. Put a cursor inside brackets, click on **Dynamic content**, select **value**, and then click **OK**, as shown in the following screenshot:

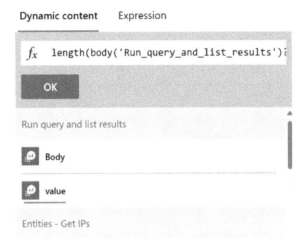

Figure 8.8 – Utilizing expressions with dynamic content

   II. For the second value, choose **is greater than** from the dropdown.

   III. For the last value, add 0 (zero).

   What this will do is check whether the watchlist query contains an IP (greater than zero) or not.

Step E – block the user if the IP is in the watchlist

In the **True** stream, we now need to configure blocking the user in the incident:

1. In the **True** stream, select **Add an action**, search for `Control`, and select the **For each** action. Let's rename it **For each – Account** by clicking on the three dots in the action name and selecting **Rename**, as shown in the following screenshot:

Figure 8.9 – The Rename action in Logic App Consumption

2. Then, in our output field, add **Accounts** from the dynamic content, as shown in the following screenshot:

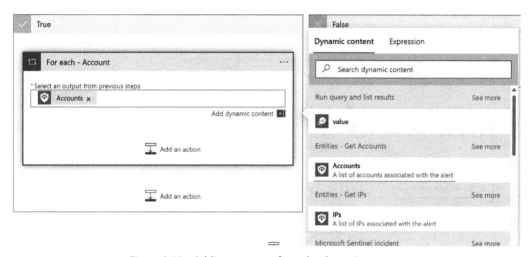

Figure 8.10 – Adding accounts from the dynamic content

3. Select **Add an action** under **For each**, search for Azure AD, and under actions, select **Update user**. The only way to authenticate this connector is by using a user identity. Click on **Sign in** and complete the sign-in process with the user that has permission to block the user in Azure AD (for example, the Global Administrator or User Administrator). Once it is authenticated, let's add the necessary fields.

4. In the **User Id or Principal Name** field, select **Expressions** and add this value:

```
concat(items('For_each__-_Account')?['Name'], '@', items('For_
each__-_Account')?['UPNSuffix'])
```

This will combine the **Name** and **UPNSuffix** values we configured in **Entity mapping** and create a full user principal name that we will use to block the user.

5. Select **Add new parameter**, and then **Account enabled**. For its value, choose **No**. *Figure 8.11* shows the preceding steps:

Figure 8.11 – Updating the user action details

6. Let's now also add comments to keep track in the incident about the user being blocked.

7. Select **Add an action** after **Update user**, search for Microsoft Sentinel, and then select **Add comment to incident**. Rename it **Add comment to incident – user blocked**. Add the incident **ARM ID** from the dynamic content, and for a comment, add **User blocked as it signed in from malicious IP**.

8. After **User**, add the concat expression:

```
concat(items('For_each__-_Account')?['Name'], '@', items('For_
each__-_Account')?['UPNSuffix'])
```

9.   After **IP**, add **IPs Address** from the dynamic content, as shown in the following screenshot:

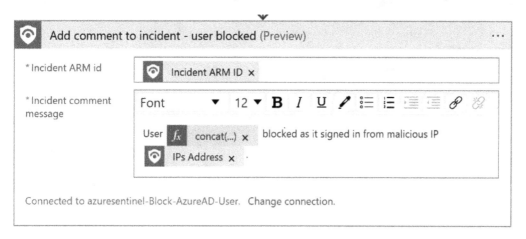

Figure 8.12 – Adding a comment to incident action details

Step F – assign permissions to a managed identity

To finish, we need to assign the **Microsoft Sentinel Responder** role to our managed identity, as we covered in *Chapter 6* and *Chapter 7*.

Our `Blok-AzureAD-User` playbook is now finished! Let's test it out!

Testing a playbook

In the next few steps, we will test our playbook and make sure that all elements work as expected:

1.   Go to **Microsoft Sentinel | Analytics** and enable and disable the **Test – Suspicious sign-in** detection rule.

2.   Go to the **Incident** tab and select the **Test – Suspicious sign-in** incident. On the right side, in summary, we can see alert enrichment about the IP address we used in *Chapter 6*. We may want to isolate the user when we see that it is malicious. Select **Actions | Run playbook**. Select **Run** next to the `Blok-AzureAD-User` playbook.

3.   Let's go back to our incident. Select **View full details** from the top tab, and then **Comments**. We can see in our comment that the user is isolated, as shown in the following screenshot:

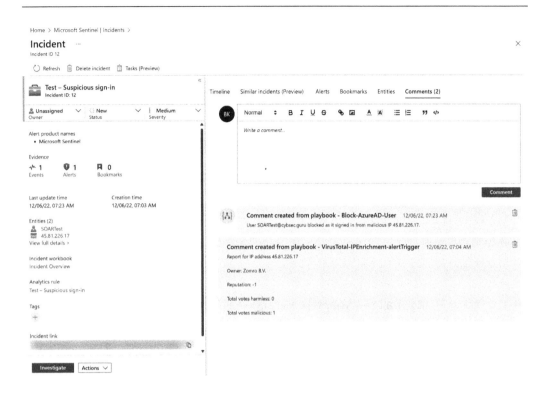

Figure 8.13 – A comment added after a successful playbook run

4.   If we go to the Azure AD portal and select **SOAR Test user** under **Users**, we will see that the user account is disabled, as shown in the following screenshot:

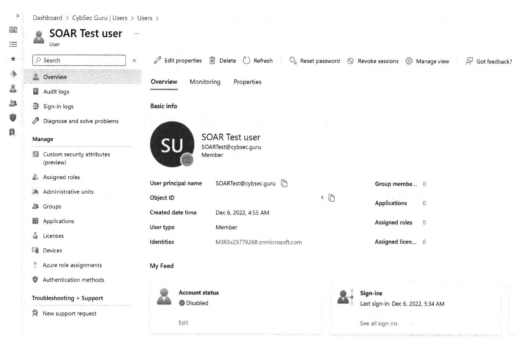

Figure 8.14 – The account is marked Disabled in Azure AD

5. If we try to log in with this user using the **InPrivate** tab again, we will get the notification that the user is locked, as shown in the following screenshot:

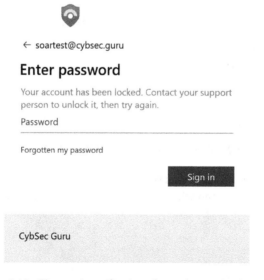

Figure 8.15 – The user's notification after trying to sign in again

Usually, we wouldn't create this playbook without an additional step, such as the process of approval or at least checking the VIP watchlist so that we do not block users that we shouldn't. In many cases, it will depend on an organization's policy. What I would typically do in this case is make sure to notify the manager that the user is blocked because of the security incident, but to do so, we need to add manager details to Azure AD and make sure that it is updated. This is not important for our exercise, but it is really important in a real-life scenario when utilizing an incident or alert trigger. By adding this step, it will be easier to deal with the upcoming entity trigger as we will have the option to run the playbook on a single entity. With this, we have successfully created and tested a playbook to block an Azure AD user. In the next hands-on example, we will be creating a playbook to isolate a host.

Isolating a machine upon new malware detection

In this example, we will use integration between **Microsoft Sentinel** and **Microsoft 365 Defender** to run a playbook and isolate a machine infected by ransomware.

Before we begin, you will need the following:

- You need to have access to Microsoft Sentinel with appropriate permissions (Microsoft Sentinel Contributor, Logic App Contributor, and permission to assign RBAC controls – Owner or User Access Administrator)
- Global Administrator or Security Administrator role permissions in Azure AD

Creating a playbook

We will need to do the following in this example:

1. Gain access to **Microsoft Defender for Endpoint** (**MDE**).
2. Connect MDE to Microsoft Sentinel for incident synchronization.
3. Create a test alert using MDE.
4. Create a playbook to isolate a machine in MDE and assign permissions.

Step 1 – gain access to MDE

The first step is to have access to MDE. If you already have MDE on your tenant, you can skip to *step 2*:

1. Go to `admin.microsoft.com` from your tenant, and go to **Billing** | **Purchase services**. Search for `Microsoft 365 E5` and select **Details** next to it, as shown in the following screenshot:

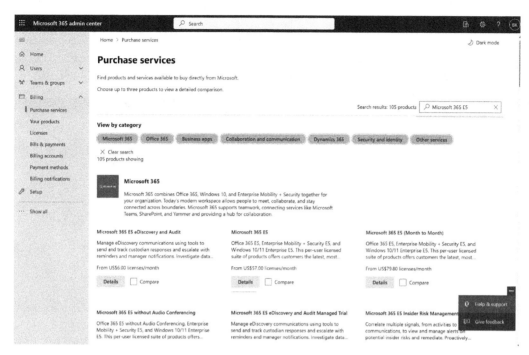

Figure 8.16 – Adding a trial in Microsoft 365

2. On the **Product details** page, select **Start free trial**, as shown in the following screenshot:

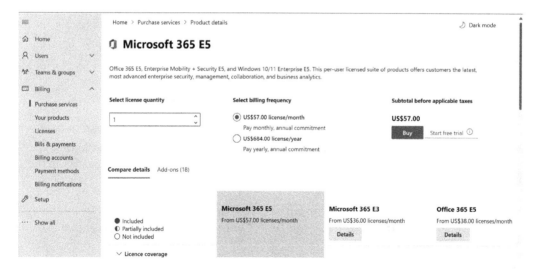

Figure 8.17 – Starting a free trial

3. Follow the steps to prove you are not a robot and opt to start a free trial. In the next window, fill in any requirements and select **Try now**. This will activate your trial license for 30 days. It will take 5–10 minutes to initialize your environment; I usually go to security.microsoft. com and select **Devices**. Once you get to the screen shown in *Figure 8.18*, proceed to *step 2*.

Figure 8.18 – The Devices view in the Microsoft 365 Security portal

Now that we have MDE, we can connect it to Microsoft Sentinel.

Step 2 – connect MDE to Microsoft Sentinel for incident synchronization

Step 2 is to connect bidirectional synchronization of incidents between **Microsoft 365 Defender** (**M365D**), which includes MDE, and Microsoft Sentinel. This will sync incidents. Furthermore, if they are closed in either Microsoft Sentinel or M365D, they will also be closed simultaneously in the other platform:

1. Go to **Microsoft Sentinel** | **Data connectors**, search for Microsoft 365 Defender, and then select the **Open connector** page. Select **Connect incidents & alerts**, as shown in the following screenshot:

Figure 8.19 – Connect the Microsoft 365 Defender data connector

With this configuration, we have connected M365D, and with it MDE, to Microsoft Sentinel.

Step 3 – create a test alert using MDE

Since we will need an incident to run our playbook during its creation, let's create a sample one:

1. Go to `security.microsoft.com` and, under **Endpoints | Evaluation & tutorials**, select **Evaluation lab**. Then, select **Setup lab**, as shown in the following screenshot:

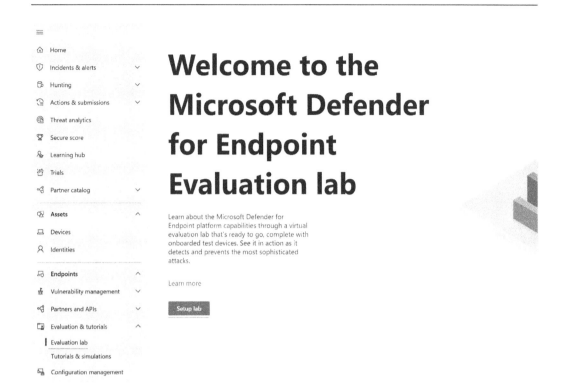

Figure 8.20 – Microsoft Defender for Endpoint Evaluation lab

2.   Choose three devices and click **Next**. Under **Install simulators agent**, check the boxes under **Microsoft privacy statement** and **Select vendors**, as shown in the following screenshot:

Lab configuration

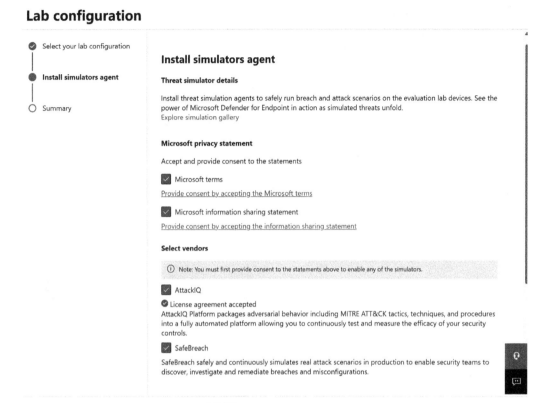

Figure 8.21 – Evaluation lab configuration

3. Select **Next** and then **Setup lab**.

4. Once our evaluation lab is deployed, we first need to add a device by selecting **Add device** and selecting the device type. We will use Windows 11 and then select **Add device**, as shown in the following screenshot:

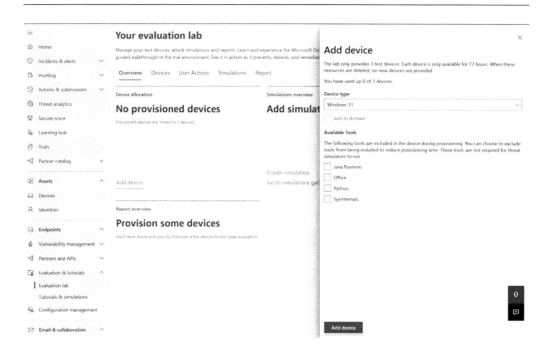

Figure 8.22 – Adding a test device to the evaluation lab

5. You will get the device name, user name, and password so that we can connect to the device. We will need it later to verify device isolation.

    We will need to wait until the device is provisioned for us, as in *Figure 8.23*.

Device allocation

1 active device

Only 3 test devices are provided. Once provisioned, it is only available for 72 hours. Depending on your monthly allotted resource consumption, you may be able to request for more devices.

testmachine1   **71/72**

View full list

Figure 8.23 – Device successfully added

6. Once the device is provisioned, select **Create simulation**. For **Select simulator**, select **SafeBreach**, and for **Select simulation**, choose **Known Ransomware Infection**. In **Select device**, select our created device. It should be **testmachine1**. Select **Create simulation**, as shown in the following screenshot:

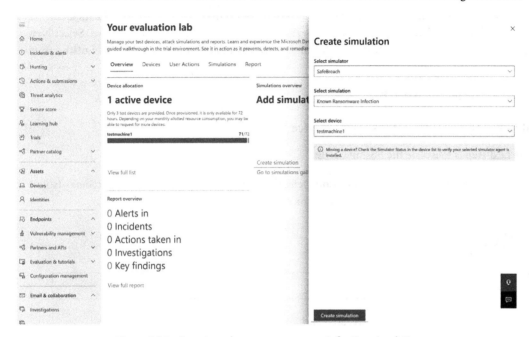

Figure 8.24 – Running a known ransomware infection simulation

While we wait for that simulation to complete and incidents to be generated, let's go and start the playbook creation.

Step 4 – create a playbook to isolate a machine in MDE

In step 4, we need to create a playbook and assign appropriate permissions to be able to isolate a device. We will follow these steps in the process:

- Step A – create a playbook with incident trigger and extract host entities
- Step B – add an action to isolate a device
- Step C – add a comment to the incident action and assign permission to a managed identity

Let's start creating the playbook.

Step A – create a playbook with incident trigger and extract host entities

To start, we need to create a playbook with an incident trigger and add the **Entities - Get Hosts** action:

1. Go to **Microsoft Sentinel | Automation | Create | Playbook with incident trigger**.

2. Select the subscription and resource group, and for the playbook name, use `Isolate-MDE-Machine`. In **Connections**, leave the managed identity and continue to **Review and create**. Select **Create and continue to designer**.

3. Our trigger is already added, and we need to add an action to get hosts from the incident. We will use a similar action as we used for IPs and accounts, but this one is for hosts. Select **New step**, search for `Microsoft Sentinel`, and select **Entities - Get Hosts**. Add **Entities** from dynamic content.

Now we can proceed to step B.

Step B – add an action to isolate a device

The next step is to isolate the host in MDE:

1. Select **New step**, search for **Microsoft Defender ATP**, and from **Actions**, select **Actions - Isolate machine** as shown in the following screenshot:

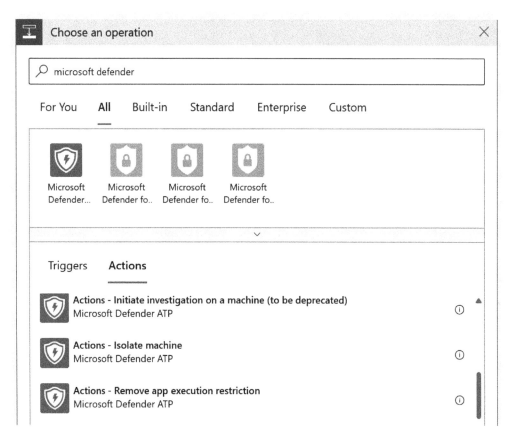

Figure 8.25 – MDE (Microsoft Defender ATP) actions

Note that MDE was previously called Microsoft Defender **Advanced Threat Protection** (**ATP**) and this is still visible in some places.

2. We need to authenticate the connection, and we can choose between managed identity, service principal, or user identity. If we are utilizing a managed identity or service principal, we need to assign appropriate permissions – either an API permission or an Azure AD role such as **Security administrator**. In this case, we will utilize user identity, and this user needs to have a **Global administrator** or **Security administrator** permission.

3. Select **Sign in** and finish signing in to authenticate the connector.

4. Now we can see that we need **Machine ID** to connect and that data is unavailable from dynamic content. We will need to run the playbook without this action to see where this specific information is located. Let's delete this action (the same way as renaming but choosing to delete) and save our playbook.

5. In a new browser tab, open **Microsoft Sentinel | Incidents**. We should see the incident we generated in the M365D portal. If it is not there yet, give it a couple more minutes.

6. Once incidents appear, choose one and run our playbook on it.

7. Go back to our playbook, select **Overview**, and then **Succeeded**, as in *Figure 8.26*.

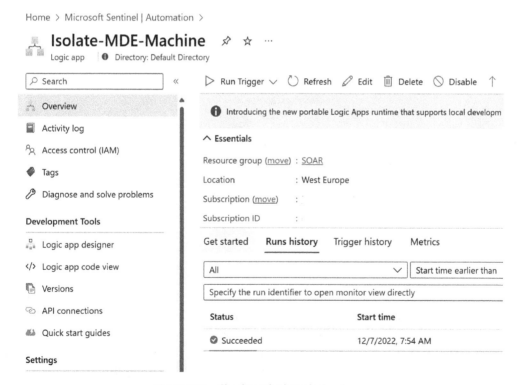

Figure 8.26 – Checking the last playbook run

8. We will see the details of the run. We will expand **Entities - Get Hosts** and check **Outputs** from it. I usually select **Show raw output** to see the full details. In it, we can see that it contains the **MdatpDeviceId** field that we need, as shown in the following figure:

Outputs   ...
Entities - Get Hosts

```
{
    "statusCode": 200,
    "headers": {
        "Pragma": "no-cache",
        "Transfer-Encoding": "chunked",
        "Vary": "Accept-Encoding",
        "Cache-Control": "no-store, no-cache",
        "Set-Cookie": "A.                                                    ,HttpOnly;
        "x-ms-request-id": "b0.                        /",
        "Strict-Transport-Security": "                        is",
        "X-Content-Type-Options": "nosniff",
        "X-Frame-Options": "DENY",
        "Timing-Allow-Origin": "*",
        "x-ms-apihub-cached-response": "false",
        "x-ms-apihub-obo": "false",
        "Date": "Wed, 07 Dec 2022 07:54:38 GMT",
        "Content-Type": "application/json; charset=utf-8",
        "Expires": "-1",
        "Content-Length": "516"
    },
    "body": {
        "Hosts": [
            {
                "hostName": "testmachine1",
                "osFamily": "Windows",
                "osVersion": "21H2",
                "additionalData": {
                    "MdatpDeviceId": "c27a373660a3b534c032d70f204067c49d793e54",
                    "FQDN": "testmachine1",
                    "RiskScore": "High",
                    "HealthStatus": "Active",
                    "LastSeen": "2022-12-07T07:27:11.2020735Z",
                    "LastExternalIpAddress": "13.74.18.134",
                    "LastIpAddress": "10.1.1.68",
                    "AvStatus": "Unknown",
                    "OnboardingStatus": "Onboarded",
                    "LoggedOnUsers": "[{\"AccountName\":\"administrator1\",\"DomainName\":\"TestMachine1\"}]"
                },
                "friendlyName": "testmachine1",
                "Type": "host"
```

Figure 8.27 – Showing raw outputs for more details

9. But how do we add it to our playbook? We will use expressions to get this info, and it is important to know where it is located. We need to remember the path – **body** | **Hosts** | **additionalData** | **MdatpDeviceId** – as shown in the following screenshot:

```
"body": {
    "Hosts": [
        {
            "hostName": "testmachine1",
            "osFamily": "Windows",
            "osVersion": "21H2",
            "additionalData": {
                "MdatpDeviceId": "c27a373660a3b534c032d70f204067c49d793e54",
                "FQDN": "testmachine1",
                "RiskScore": "High",
                "HealthStatus": "Active",
                "LastSeen": "2022-12-07T07:27:11.2020735Z",
                "LastExternalIpAddress": "13.74.18.134",
                "LastIpAddress": "10.1.1.68",
                "AvStatus": "Unknown",
                "OnboardingStatus": "Onboarded",
                "LoggedOnUsers": "[{\"AccountName\":\"administrator1\",\"DomainNa
            },
            "friendlyName": "testmachine1",
```

Figure 8.28 – The path to MdatpDeviceId

10. Let's go back to the designer of our playbook. Now, we need to add an **MdatpDeviceId** field, and as we will add it as an expression, we will need to first add a **For each** loop, as we can have more than one host in the incident.

11. Select **New step**, and then **Control | For each**. For **Output**, choose **Hosts** from the dynamic content. In the **For each** loop, select **Add an action**, and search for and select **Microsoft Defender ATP | Actions – Isolate machine**. To add **Machine ID**, click inside the empty space, select **Expression**, and add the following:

```
items('For_each')?['additionalData']?['MdatpDeviceId']
```

In the comment space, add **Isolated using the Isolate-MDE-Machine playbook**.

Figure 8.29 shows this configuration.

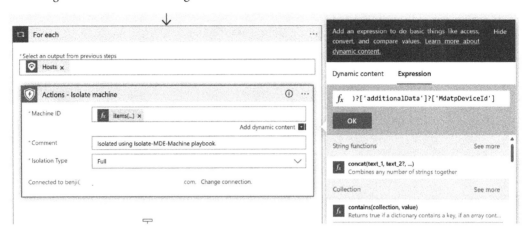

Figure 8.29 – Using Expression to write the path

But why do we follow the `['additionalData']?['MdatpDeviceId']` path in step 11, and not the full `body('Entities_-_Get_Hosts')?['Hosts']?['additionalData']?['MdatpDeviceId']` path? It is because in `For each Hosts`, we already follow the `body('Entities_-_Get_Hosts')?['Hosts']` path. So, in this field, we can use `items('For_each')` and add the rest of the path.

If **Hosts** was not an array field, we could use the only full path without using **For each**.

Now we can proceed to step C.

Step C – add a comment to the incident action and assign permission to a managed identity

In step C, we need to add an action to add a comment to the incident stating which host is isolated and assign permissions to the managed identity:

1. We will now also add a comment to the incident. Select **Add an action** | **Microsoft Sentinel** | **Add comment to incident**. Add **Incident ARM ID** from dynamic content.

    In the comment message, add the following:

    `Machine is isolated.`

    And then, after the machine, add one more expression. Instead of `['additionalData']?['MdatpDeviceId']`, we will use `['friendlyName']`:

    `items('For_each')?['friendlyName']`

Figure 8.30 shows this configuration.

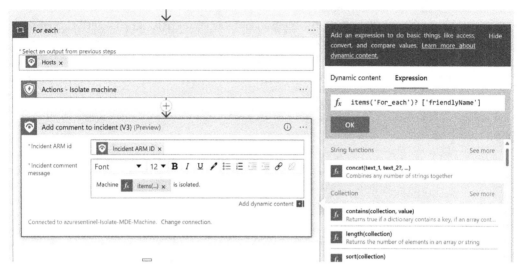

Figure 8.30 – Add friendlyName from the MDE action JSON response

2. Save the playbook and assign the **Microsoft Sentinel Responder** role to the managed identity.

Once our playbook is saved, we can test it on the test incident we created.

Testing a playbook

The last part, as in previous hands-on examples, is to test our newly created playbook:

1. Let's first connect to our device to see device isolation. Go to the MDE **Evaluation lab** space, and select the **Devices** tab. And under **Actions**, select **Connect**, as demonstrated in *Figure 8.31*. Use the password you have or generate a new password.

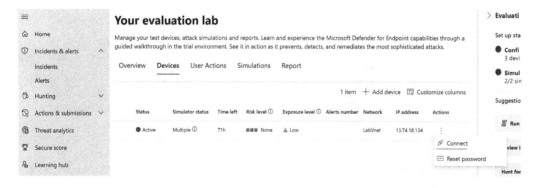

Figure 8.31 – Connecting to the test machine

2. Once logged in, go to **Microsoft Sentinel | Incidents** and run the playbook on one of the MDE incidents. You will probably be unable to see incidents, as they will be remediated from the MDE side using the **Automated Investigation and Response (AIR)** functionality. Change the incident page view to see closed incidents as well and run the playbook. The incident name should be **Multiple threat families detected including Ransomware on one endpoint**, and under **Product names**, you can see **Microsoft 365 Defender**, as shown in the following screenshot:

Figure 8.32 – Running the playbook from Microsoft Sentinel

3. Let's go back to the virtual machine we connected to; after a few seconds, we will lose our connection, as we can see in the following screenshot:

Figure 8.33 – A lost connection to the test machine

4. If we go to **M365D portal | Devices | testmachine1**, we will see that the device is isolated and we can now release it from isolation. If we release it from isolation, we can connect to the machine again, as shown in the following screenshot:

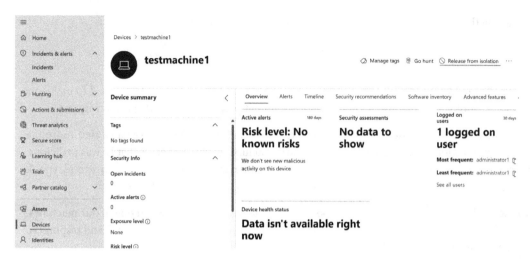

Figure 8.34 – Notification in the Devices tab in the Microsoft 365 Security portal

5. In Microsoft Sentinel, we can see our comment, as shown in the following screenshot:

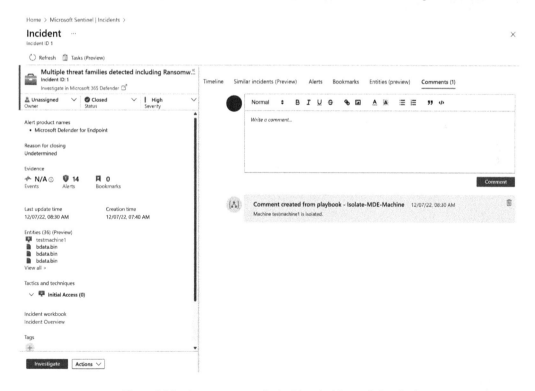

Figure 8.35 – A comment on the incident in Microsoft Sentinel

By testing the playbook, we have finished our last hands-on example.

Summary

In this chapter, we continued with hands-on examples and focused on how to respond to incidents using playbooks, and how we can utilize our enrichment playbooks to decide how to respond.

In the first example, we used a playbook to block a user in Azure AD when we had a sign-in from a suspicious location, and we used our **MaliciousIP** watchlist to check whether that IP was malicious. Plus, we had a comment from the `VirusTotal-IPEnrichment-alertTrigger` playbook (which we created in *Chapter 6*) to get information from VirusTotal as well.

The second example focused on our response when malware was detected on one of our machines. We utilized MDE and its evaluation lab to generate incidents, sync them to Sentinel, and respond to isolate the machine in MDE itself.

In our final chapter, we will provide a few important tips and tricks for working with Microsoft Sentinel playbooks. If you decide to go even deeper into Microsoft Sentinel playbooks, these tips and tricks will save you time and provide more insight into your day-to-day playbook operations.

Mastering Microsoft Sentinel Automation: Tips and Tricks

In the previous three chapters, we focused on hands-on examples that hopefully helped you understand how to utilize automation.

We started with examples of how to enrich incidents on the incident creation for faster triage in *Chapter 6* and continued with examples of how to utilize automation to manage incidents in *Chapter 7*. In the previous chapter, we focused on how to utilize automation to respond to incidents. We utilized the two most common response automation techniques – block the user and isolate a device.

This chapter will focus on tips and tricks for mastering automation when working with Microsoft Sentinel playbooks.

In this final chapter of the book, we will go through the following topics:

- Best practices for working with dynamic content and expressions
- Understanding the **HTTP** action and its usage
- Exploring more playbook actions

Let's get into it!

Best practices for working with dynamic content and expressions

In the examples discussed in previous chapters, we used dynamic content over static content wherever we could. This is because with dynamic content, we receive the latest data to make appropriate decisions.

We learned that all actions in the playbooks that speak with external services, such as Microsoft Sentinel, Azure AD, MDE, and VirusTotal, use API calls. All APIs will return responses in JSON format, as we saw in our MDE example in *Chapter 8* when we searched for MdatpDeviceId. While most of the actions map API responses to dynamic content, not all response fields from the API are available as dynamic content, and we saw that with `MdatpDeviceId`.

In this case, when dynamic content is not available for the specific field we need, we can utilize expressions instead of dynamic content to get the data inside the playbook – but how do we utilize expressions?

Let's open the **Isolate-MDE-Machine** playbook and expand **For each** and **Actions – Isolate machine**. We can see that our expressions changed the look. Instead of the `items('For_each')?['additionalData']?['MdatpDeviceId']` expression, we can see `additionalData.MdatpDeviceId` as dynamic content. If we search for it in dynamic content, we cannot find it. So, how come we can see it as dynamic content when it is unavailable?

Dynamic content is nothing but a Logic Apps expression created to aid the understanding of the user. If we copy `additionalData.MdatpDeviceId` and paste it in the text editor, we can see our expression with an added `@{expression}` part, which Logic Apps adds automatically for dynamic content.

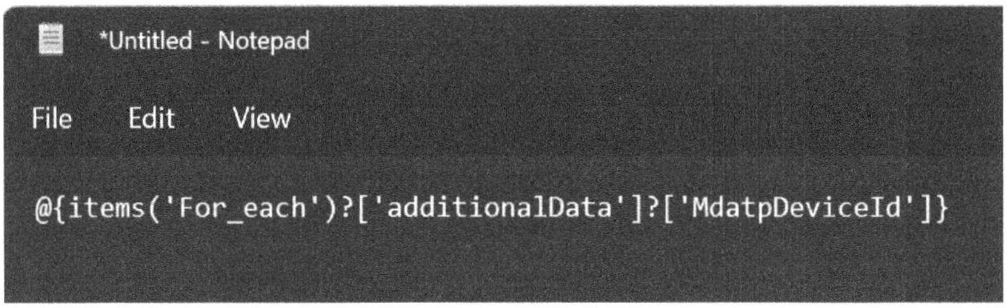

Figure 9.1 – MdatpDeviceId expression example

We can see the same syntax if we copy the `Hosts` content we added as dynamic content in the same playbook.

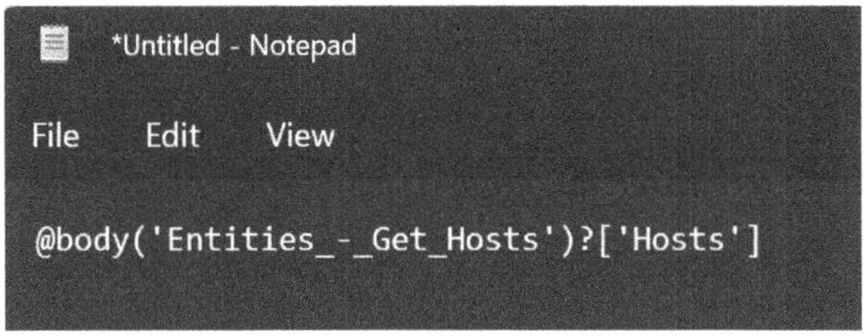

Figure 9.2 – Hosts dynamic content expression example

If we open our successful run from **Overview** and select **Show raw outputs**, we can see the path for the `Hosts` array:

```
"body": {
    "Hosts": [
        {
            "hostName": "testmachine1",
            "osFamily": "Windows",
            "osVersion": "21H2",
            "additionalData": {
                "MdatpDeviceId": "c27a373660a3b534c032d70f204067c49d793e54",
                "FQDN": "testmachine1",
                "RiskScore": "None",
                "HealthStatus": "Active",
                "LastSeen": "2022-12-07T07:27:11.2020735Z",
                "LastExternalIpAddress": "13.74.18.134",
                "LastIpAddress": "10.1.1.68",
                "AvStatus": "Unknown",
                "OnboardingStatus": "Onboarded",
                "LoggedOnUsers": "[{\"AccountName\":\"administrator1\",\"DomainName\":\"TestMachine1\"}]"
            },
            "friendlyName": "testmachine1",
```

Figure 9.3 – Location of Hosts in the JSON output

If we need to check this for any other action, it will look like `body('name_of_the_action')?['next_step_in_JSON']?['next_step_in_JSON']`.

It is important to note here that for the name of the action, where we have white space, we need to replace it with _.

In our example with `MdatpDeviceId`, we followed a path of `body` | `Hosts` | `additionalData` | `MdatpDeviceId`.

The second example expression we used in the **Isolate-MDE-Machine** playbook covered in *Chapter 8, Respond to Incidents Using Automation*, was to add `friendlyName` to the **Add comment to incident** action. As dynamic content, it was not available, but it was available as a JSON response. So, we used the expression and added `items('For_each')?['friendlyName']`.

Can you utilize this for any response? Absolutely. If you receive data in JSON, you can extract it using expressions.

Let's go through some expressions you will benefit from with Microsoft Sentinel playbooks:

- `length()` is one more that we used in our examples to see whether the **Azure Monitor Logs** action returned results or not. So, if the action is empty, the `length()` result will be 0. How it works is if we add `length('SOAR')`, 4 will be returned (the number of characters in the string), and if it's an array, such as `length('SOAR','is','the','best',', period!')`, 5 will be returned. In **Azure Monitor Logs**, if we don't rename the action, the syntax will be as follows:

    ```
    length(body('Run_query_and_list_results')?['value'])
    ```

- `first()` and `last()` will return the first or last letter in the string or the first or last entry in the array. Let's see some examples:
  - For `first('SOAR')`, the result is S, while for `last('SOAR')`, the result is R
  - For `first('SOAR','is','the','best',',period!')`, the result is SOAR, while for `last('SOAR','is','the','best',',period!')`, the result is ,period!

  If we have multiple results in our **Run query and list results** action in **Azure Monitor Logs** and we only want to see the first one, the expression would be as follows:

  ```
  first(body('Run_query_and_list_results')?['value'])
  ```

- `join()` will return array values in a single string. This expression must have two entries – an array and a delimiter that will separate each array value.

  In our **Isolate-MDE-Machine** playbook, let's say we grabbed `friendlyName` and added it to the incident. As `friendlyName` is part of the array, it is in the `for each` loop, but if we want to make a list of all `friendlyName` fields when we have more than one host in the incident, we can use `join()` for it. The expression would look like this:

  ```
  join(body('Entities_-_Get_Hosts')?['Hosts']?[' friendlyName'],
  ', ')
  ```

  The result would be hostname 1, hostname 2, hostname 3, and so on.

- `empty()` will return `true` or `false` depending on whether the string or array is empty. This can be used similarly to `length()` to see whether we have a result or not.

  An expression equivalent to `length(body('Run_query_and_list_results')?['value'])` would be `empty(body('Run_query_and_list_results')?['value'])`. For the **Condition** action, we would use `'is equal to'` and then use `true` or `false`.

- `concat()` is similar to `join()`, but here, we can combine two or more strings into one string. We utilize it in the **Block-AzureAD-User** playbook to join `Name` and `UPNSuffix` to create the user's principal name, which is needed to block the user.

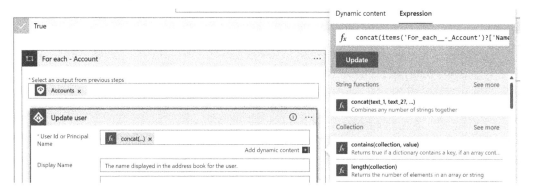

Figure 9.4 – concat() expression

The expression we used is as follows:

```
concat(items('For_each_-_Account')?['Name'], '@', items('For_
each_-_Account')?['UPNSuffix'])
```

- `split()` is used when we have multiple entries in one string and we want to separate them. The syntax requires a string and a common delimiter that is used to separate values. Here's an example:

```
split('Incident 1;Incident 2;Incident 3', ';')
```

– result will be `["Incident 1", "Incident 2", "Incident 3"]`

There is one action in Logic Apps that I personally like more than any other. I call it a DIY action, but the official name is the **HTTP action**.

Understanding the HTTP action and its usage

As already mentioned, all the Logic Apps actions that Microsoft Sentinel uses in playbooks are actually API calls. Native actions are just represented more nicely as part of a GUI, making it easier for users to utilize them. Adding dynamic content is much easier than writing the body of an API call in JSON.

But why, then, do I like the **HTTP** action more than any other? Because it allows us to create actions per our own needs, and we can also utilize different authentication methods.

Elements of the HTTP action

Firstly, what are the main elements of the **HTTP** action? Let's look at them in the following list:

- **Method**: This states the API method. The most popular are GET, POST, PATCH, PUT, and DELETE.
- **URI**: The **Uniform Resource Identifier** (**URI**) is the API call itself. It normally looks like a URL.
- **Headers**: This defines whether an API call needs to have any headers.
- **Queries**: This defines whether the API call needs to have any queries.
- **Body**: Here, we will typically add a JSON request.
- **Cookie**: Here, we enter an HTTP cookie.
- **Authentication**: This is a way to authenticate API calls. Examples are a username and password, client certificate, Active Directory OAuth authentication (with a service principal), raw authentication, and managed identity.

The preceding elements of the **HTTP** action can be seen in the following screenshot:

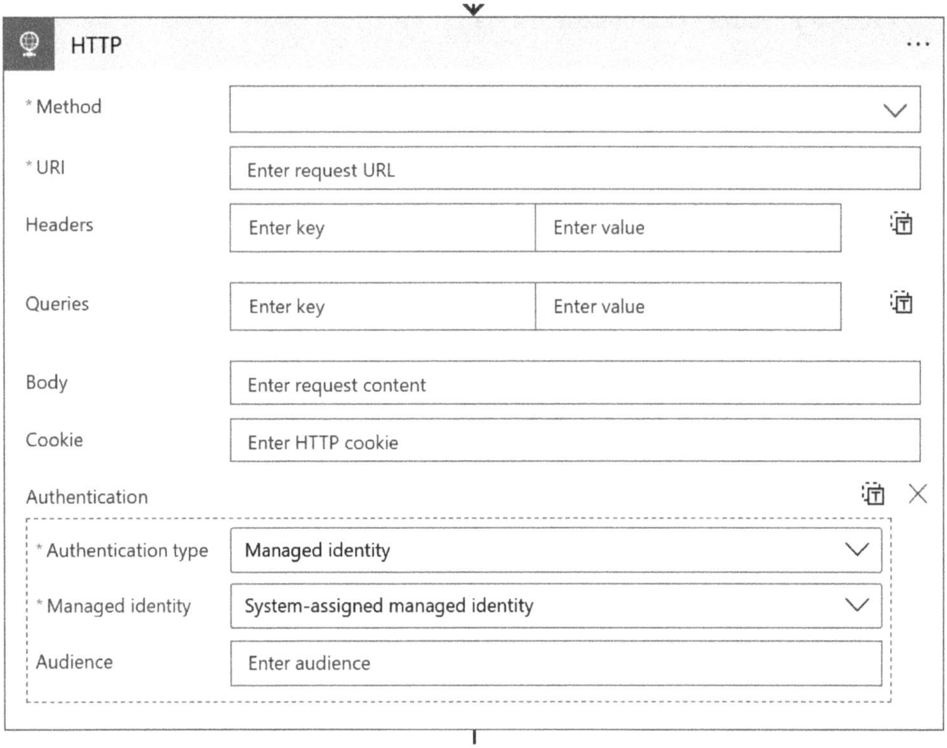

Figure 9.5 – HTTP action in Logic Apps

I usually use the **HTTP** action when I don't have managed identity or service principal support and I don't want to use user identity for authentication. One example where we have to use user identity for authentication is the **Azure AD** connection.

For example, to block a user from signing in, in the **Block-AzureAD-User** playbook, we can use user identity to authenticate the connection. The user must always have permission to perform block user action when the playbook runs. The action will fail if the user who authenticated the playbook doesn't have permission. I'm someone who tries to apply the principle of least privilege so that users have high levels of privilege only when needed. Because of that, the Azure AD Logic Apps connection is something I use to make up for a lack of managed identity and/or service principal support. Many companies also utilize **Privileged Identity Management** (**PIM**) with Azure AD so that users have privileges only when needed and not always.

How do we utilize the **HTTP** action? Is it hard?

Utilizing the HTTP action

Once you have used the **HTTP** action a few times, you'll realize it's easier than advertised. We first need to know what API calls we need to use. Since we need an API call to Azure AD, we will search for the Azure AD API under Microsoft Graph. We will search for how to update user details, and we can do so using an update user call. The URL we will use is `https://learn.microsoft.com/en-us/graph/api/user-update?view=graph-rest-1.0&tabs=http`. Here, we need to locate the details we will use in our **HTTP** action, so let's get started:

1. We need a method. We can see that we need to utilize `PATCH`.

```
HTTP

PATCH /users/{id | userPrincipalName}
```

Figure 9.6 – The API method

2. We can see in the extension after `PATCH` that the URI will be `/users/{id | userPrincipalName}`. Is that the whole URI? No. Scroll to one example in the link mentioned previously to see the whole URI or read the introduction details for the API.

```
HTTP                                                    Copy

PATCH https://graph.microsoft.com/v1.0/me
Content-type: application/json

{
```

Figure 9.7 – The API URI

We will use the correct user principal name and not `me`.

3. Do we need headers or queries? We will see that we need two headers – **Authorization** and **Content-Type**. The authorization will take the form of a service principal or managed identity, but we can also generate tokens and utilize them in the header.

Header	Value
Authorization	Bearer {token}. Required.
Content-Type	application/json

Figure 9.8 – API headers

4. What do we need in the body? First, we must ensure that we don't have mandatory body fields. A quick look will show that there is no indication of any, and there is a notification saying that any value from the body that is not included will contain its previous value. We need to block the user from signing in. For that, we will look at the **accountEnabled** body value. We must also take note of what type it is (a string, array, Boolean, and so on) so that we know what kind of entry we need. For **accountEnabled**, we can see that the value type is **Boolean**, which means that entry values are either `true` or `false`.

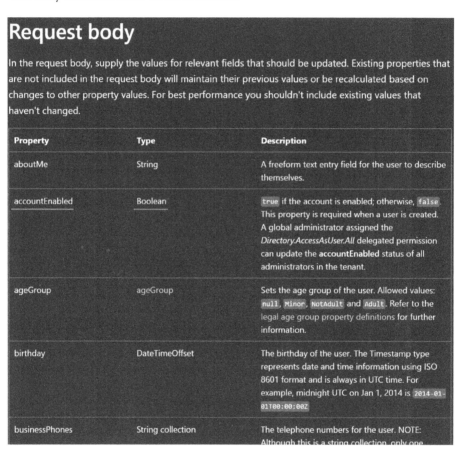

Request body

In the request body, supply the values for relevant fields that should be updated. Existing properties that are not included in the request body will maintain their previous values or be recalculated based on changes to other property values. For best performance you shouldn't include existing values that haven't changed.

Property	Type	Description
aboutMe	String	A freeform text entry field for the user to describe themselves.
accountEnabled	Boolean	`true` if the account is enabled; otherwise, `false`. This property is required when a user is created. A global administrator assigned the *Directory.AccessAsUser.All* delegated permission can update the **accountEnabled** status of all administrators in the tenant.
ageGroup	ageGroup	Sets the age group of the user. Allowed values: `null`, `Minor`, `NotAdult` and `Adult`. Refer to the legal age group property definitions for further information.
birthday	DateTimeOffset	The birthday of the user. The Timestamp type represents date and time information using ISO 8601 format and is always in UTC time. For example, midnight UTC on Jan 1, 2014 is `2014-01-01T00:00:00Z`
businessPhones	String collection	The telephone numbers for the user. NOTE: Although this is a string collection, only one

Figure 9.9 – The API body

5. Now, we also need to see what type of permissions identity is needed to perform this action. We need to look at **Application** permissions that use a service principal or managed identity. The needed permissions are `User.ReadWrite.All`, `User.ManageIdentities. All`, and `Directory.ReadWrite.All`. We can add either these API permissions or an Azure AD role that has these permissions. Examples of those Azure AD roles are **Global administrator** and **User administrator**.

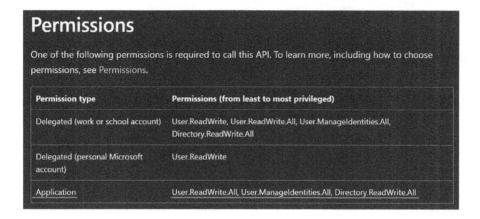

Figure 9.10 – API permissions

Those are the details we will use in our **HTTP** action, following which our HTTP call to block Azure AD users would look like this:

HTTP	···

* Method	PATCH ⌄
* URI	https://graph.microsoft.com/v1.0/ *fx* concat(...) ✕
Headers	Content-Type	application/json	✕ 🗓
	Enter key	Enter value	
Queries	Enter key	Enter value	🗓
Body	{ "accountEnabled": false }	
Cookie	Enter HTTP cookie	
Authentication		🗓 ✕
* Authentication type	Managed identity	⌄
* Managed identity	System-assigned managed identity	⌄
Audience	https://graph.microsoft.com	

Figure 9.11 – The HTTP action with details

But how do we apply these permissions?

Applying API permissions to a managed identity

If we are using a managed identity, we can assign Azure AD roles from the GUI, but to assign API permissions for the managed identity, following the principle of least privilege, we have to utilize PowerShell. We will need to connect to the Azure AD PowerShell module, sign in as a user who has permission to assign API calls, and then utilize the `Get-AzureADServicePrincipal` and `New-AzureAdServiceAppRoleAssignment` PowerShell commands. An example can be seen in the official Microsoft Sentinel GitHub repository – `https://github.com/Azure/Azure-Sentinel/tree/master/Solutions/Azure%20Active%20Directory/Playbooks/Block-AADUser`.

In the following example, we can see a shorter version, to explain segments of the code in question. In the first part of PowerShell code, we will need to add the ID of our managed identity:

```
$MIGuid = "<Enter your managed identity guid here>"
$MI = Get-AzureADServicePrincipal -ObjectId $MIGuid
```

In the next part, we will define the application ID from Azure AD (in this case, the Microsoft Graph API application ID), as well as API permissions that we need to apply to the managed identity:

```
$GraphAppId = "00000003-0000-0000-c000-000000000000"
$PermissionName1 = "User.Read.All"
$PermissionName2 = "User.ReadWrite.All"
```

In the final part of our PowerShell code, we will assign those permissions to the managed identity we defined at the start of the PowerShell code:

```
$GraphServicePrincipal = Get-AzureADServicePrincipal -Filter "appId eq
'$GraphAppId'"
$AppRole1 = $GraphServicePrincipal.AppRoles | Where-Object {$_.
Value -eq $PermissionName1 -and $_.AllowedMemberTypes -contains
"Application"}
New-AzureAdServiceAppRoleAssignment -ObjectId $MI.ObjectId
-PrincipalId $MI.ObjectId `
-ResourceId $GraphServicePrincipal.ObjectId -Id $AppRole1.Id

$AppRole2 = $GraphServicePrincipal.AppRoles | Where-Object {$_.
Value -eq $PermissionName2 -and $_.AllowedMemberTypes -contains
"Application"}
New-AzureAdServiceAppRoleAssignment -ObjectId $MI.ObjectId
-PrincipalId $MI.ObjectId `
-ResourceId $GraphServicePrincipal.ObjectId -Id $AppRole2.Id
```

If we want to assign API permission to the service principal, we can do so from the GUI for the Azure AD role and API permission.

To assign API permission to the service principal, go to **Azure Active Directory | App registrations**, and choose our service principal or create a new one. To make it more convenient and not have to go through creating a service principal again, I will choose the SOAR service principal we used to assign the **Log Analytics Reader** role in *Chapter 7, Manage Incidents with Automation*. In production, you will probably use a different service principal so that one identity doesn't have too many permissions assigned. Once opened, from the left menu, choose **API permissions** and then **Add a permission**. The API in question is **Microsoft Graph**, so we need to select it and then choose **Application permissions**.

We will then search for the API permissions we need – `User.ReadWrite.All`, `User.ManageIdentities.All`, and `Directory.ReadWrite.All` – and add them.

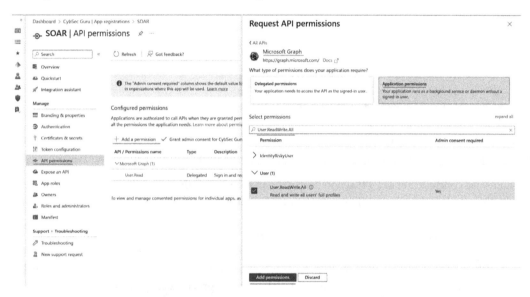

Figure 9.12 – Assigning API permissions to a service principal in Azure AD

Once we have added all permissions, we will need to add admin consent for our organization so that the specific service principal can use those API permissions.

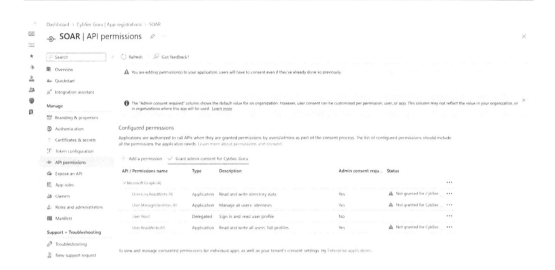

Figure 9.13 – Granting admin consent for API permissions

Once configured, we can utilize the service principal for authentication. We will need the same information as when authenticating the connection – the application ID, tenant ID, and secret.

With this, we conclude our **HTTP** action deep dive. In the next section, we will focus on a few more actions that can help us master Microsoft Sentinel playbooks.

Exploring more playbook actions

Each action in the playbook has its own settings that we can work with. We have already mentioned options to rename an action, add a comment, or delete one in previous chapters, but there are a few more great features that you need to know about:

- We have the option to configure a timeout for an action. We usually want this to be used with user input when we don't want to wait hours or days for a user to respond. Logic Apps Standard and Logic Apps Consumption can be in a running state for 90 days by default. To change it, we need to click on the three dots in the playbook and select **Settings**.

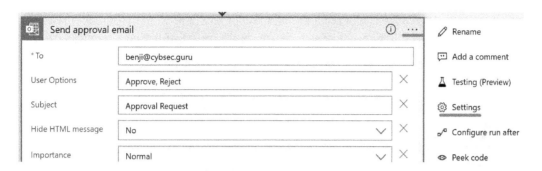

Figure 9.14 – Action settings

In the settings, we need to change the value for **Action Timeout**. It accepts input in the **ISO 8601** format. If we wanted that action to time out after one day, the input would be P1D. If we wanted it to time out in one hour, the input would be PT1H. An example of all the values needed is P1M1DT1H1M – one month, one day, one hour, one minute. You can configure years, seconds, and milliseconds, but we don't need to.

To configure the approval playbook to wait for approval for 10 minutes, we will use PT10M. Please note that P10M would be 10 months, so we need to use T for time.

Settings for 'Send approval email'

Secure Inputs
Secure inputs of the operation.
Secure Inputs
(●) Off

Secure Outputs
Secure outputs of the operation and references of output properties.
Secure Outputs
(●) Off

Action Timeout
Limit the maximum duration between the retries and asynchronous responses for this action. Note: This does not alter the request timeout of a single request.

Duration ⓘ	PT10M

Retry Policy
A retry policy applies to intermittent failures, characterized as HTTP status codes 408, 429, and 5xx, in addition to any connectivity exceptions. The default is an exponential interval policy set to retry 4 times.

Type	Default ⌄

Tracked Properties

Key	Value

Done	Cancel

Figure 9.15 – Action Timeout

If we run the playbook with this configuration and don't get a response in 10 minutes, we will see that our playbook will fail, and if we go to the run details, we can see that our timeout kicked in.

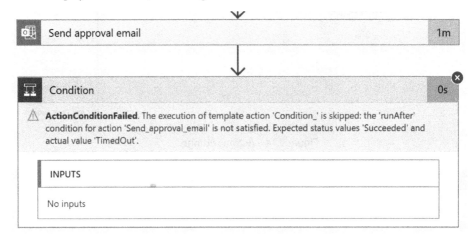

Figure 9.16 – Action timeout in the playbook run

- However, what if we don't want that playbook to fail? What if we want to continue the run and maybe add a comment for the incident that no user selection was made?

In this case, we need to go to the next action in our playbook, click the three dots, and choose **Configure run after**:

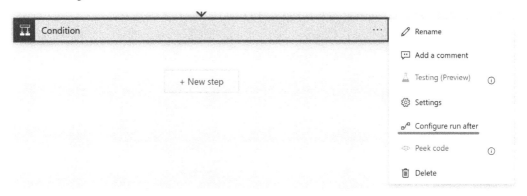

Figure 9.17 – Configure run after

In the settings, in this case, we need to check the **has timed out** checkbox as well. The same process applies if the playbook needs to keep running even if one of the actions fails. You can go to the next action and include **has failed**.

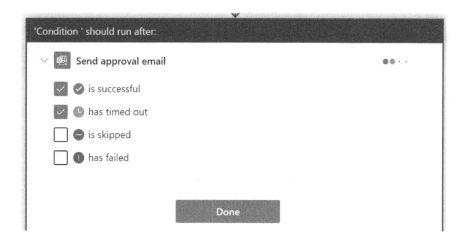

Figure 9.18 – Configure run after settings

If we rerun the playbook, we can see that the playbook didn't fail at this step.

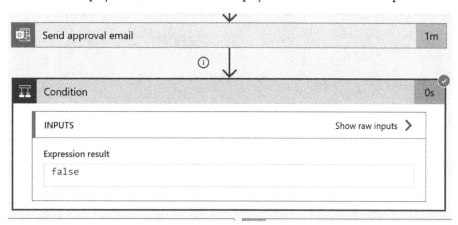

Figure 9.19 – Configuring run after for the playbook run

- The final option we will go through here is turning on secure inputs and outputs. This is one option I always use if I utilize a service principal for authentication, like with the **HTTP** action. I usually save a secret in Azure Key Vault and utilize the **Azure Key Vault** action to get the secret. Then, I turn on **Secure Inputs** and **Secure Outputs**. This will then hide secret values in the output. This is how we can hide secrets so that they cannot be accidentally shared when someone shares playbook run or action details.

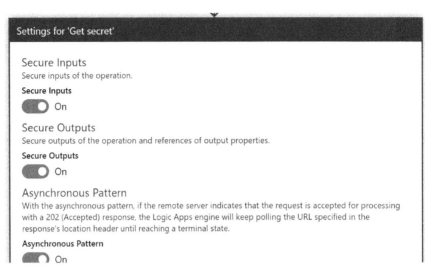

Figure 9.20 – Secure Inputs and Secure Outputs

When we run the playbook, the output in the **Get secret** action will be hidden:

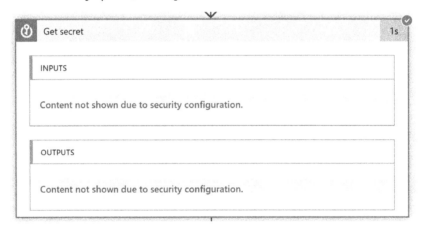

Figure 9.21 – Secure Inputs and Secure Outputs in the playbook run

Now, we will introduce a few more actions that can be useful when working with playbooks.

Switch

The **Switch** action is part of the **Control** pack, where we have already used **For Each** and **Condition**. This will allow us to select which stream of actions to take based on a specific value. This is usually when we have more than two options. In the example here, we used approval and timeout; for that, we typically have three options – **Approve**, **Reject**, and **Timeout**.

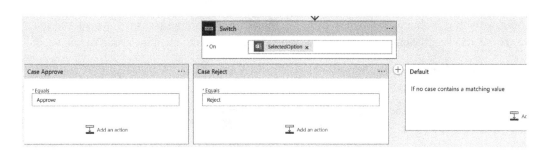

Figure 9.22 – The Switch action

Control only gives us the option of **True** or **False**, so it's probably not the most effective solution. Instead, we can use **Switch** and add as many options as we need. One of the options is the default, and we can leave it as **Timeout**, for example, and add two more options for **Approve** and **Reject**.

Select and Create HTML table

Here, I will introduce two actions that usually go together. We utilize the **Select** and **Create HTML table** actions if we want to create an HTML table based on array values, such as a list of all entities and what kinds of entities they are.

First, we will use **Select** to map columns and values. **Select** is an action located under **Data Operations**. **From** is the array from which we will extract values; in our case, it will be **Entities** from dynamic content.

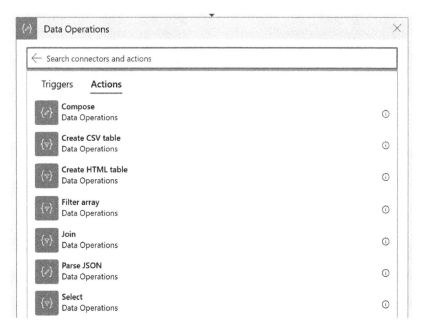

Figure 9.23 – Data Operations Logic Apps connection

In **Map**, we will add columns and values. To get the values, you can always save the playbook and run it to see the details in the output. Once we have the details, we need to fill in the action as follows:

1. In **Enter key**, we will add `Entity`, and in **Enter value**, we will add `item()?` `['properties']?['friendlyName']` from the expression.

2. In **Enter key**, we will add `Entity type`, and in **Enter value**, we will add `item()?['kind']` from the expression.

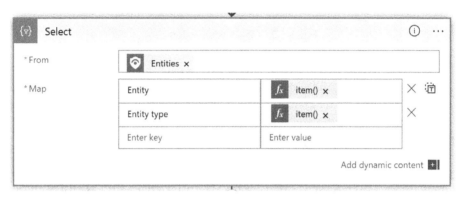

Figure 9.24 – Configuring the Select action

3. After this action, we will add the **Create HTML table** action from under **Data Operations**.

4. In **From**, we will add **Output** from the **Select** action, and we will leave **Columns** as **Automatic**.

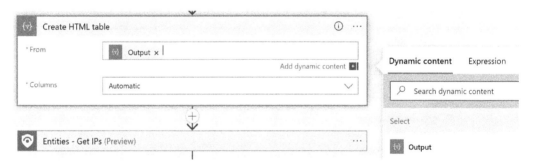

Figure 9.25 – Configuring the Create HTML table action

5. To add the HTTP table, we will add **Output** from **Create HTML table** to our comment or email action.

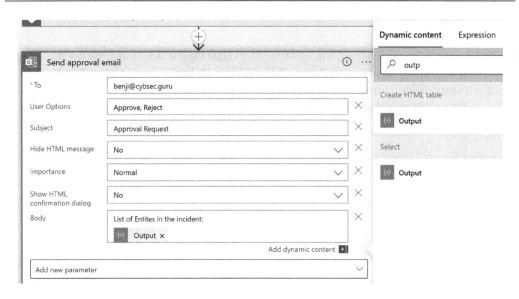

Figure 9.26 – Adding the Create HTML table output to emails

Compose

Compose is one more action available under **Data Operations** that will help you when you need to compose a text that you may want to reuse and not type all over again. You have the option to adjust the text in just one place.

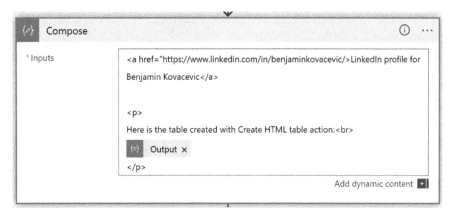

Figure 9.27 – The Compose action

If we want to format our text in HTML, for example, in **Compose**, we can format our whole email body so that it will be easier to read when we send it. We can utilize dynamic content and expression values in it as well.

Parse JSON

Parse JSON is the last example we will use, and it is part of **Data Operations** as well. We use this when we don't have dynamic content in our actions. This will typically be the case with the **HTTP** action, as it will not return fields for dynamic content. Let's use the Microsoft Graph API to get a user called SOAR test. I used **Managed identity** for authentication and was assigned a **User administrator** role.

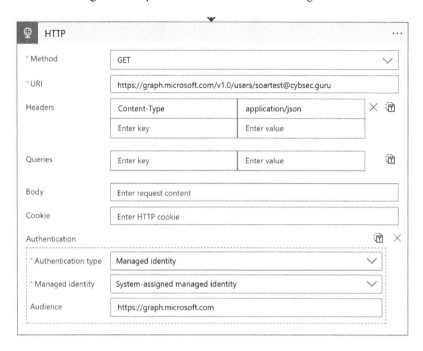

Figure 9.28 – A sample HTTP action

If I check **Dynamic content**, we can see that only the general options are there, not any fields we need.

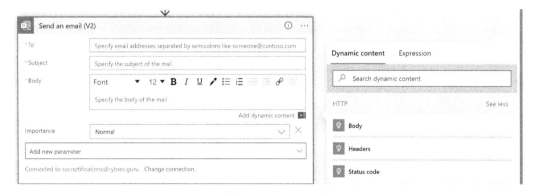

Figure 9.29 – Available dynamic content for the HTTP action

We will need to use the **Parse JSON** action to get that data from the **HTTP** action, but for that, we will need a sample schema. The easiest way to get the schema is to run a playbook with the **HTTP** action and use **Show raw output**. We will copy the body of the output, as we can choose **Body** from the **HTTP** dynamic content.

```
"body": {
    "@odata.context": "https://graph.microsoft.com/v1.0/$metadata#users/$entity",
    "businessPhones": [],
    "displayName": "SOAR Test user",
    "givenName": null,
    "jobTitle": null,
    "mail": null,
    "mobilePhone": null,
    "officeLocation": null,
    "preferredLanguage": null,
    "surname": null,
    "userPrincipalName": "SOARTest@cybsec.guru",
    "id": "46f7b9a7-0a1f-44d2-a345-47ed303b8d94"
}
```

Figure 9.30 – The raw output from the HTTP action

Let's go to **Logic Apps Designer** and add the **Parse JSON** action after the **HTTP** action. For **Content**, add **Body** from **Dynamic content**. Then, click on **Use sample payload to generate schema**.

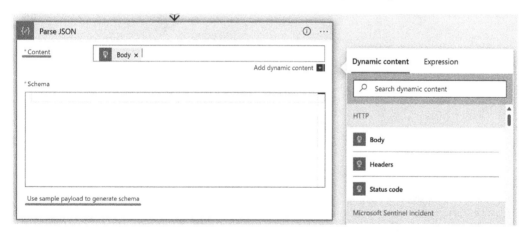

Figure 9.31 – Using a sample payload to generate a schema

Paste the raw output and select **Done**.

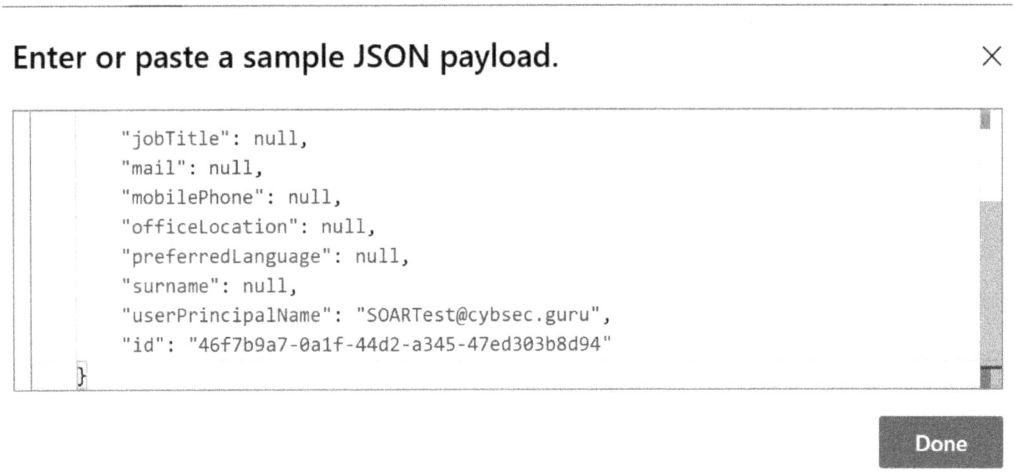

Enter or paste a sample JSON payload.   ✕

```
    "jobTitle": null,
    "mail": null,
    "mobilePhone": null,
    "officeLocation": null,
    "preferredLanguage": null,
    "surname": null,
    "userPrincipalName": "SOARTest@cybsec.guru",
    "id": "46f7b9a7-0a1f-44d2-a345-47ed303b8d94"
}
```

Done

Figure 9.32 – The raw output from the HTTP action used to generate a schema

A schema will be generated with the raw details.

{∂} **Parse JSON**   ⓘ ...

* Content

    ⏚ Body ✕

* Schema

```
{
    "type": "object",
    "properties": {
        "@@odata.context": {
            "type": "string"
        },
        "businessPhones": {
            "type": "array"
        },
```

Use sample payload to generate schema

Figure 9.33 – Schema generated by the sample JSON payload

Now, we can utilize these details as dynamic content in the rest of the playbook.

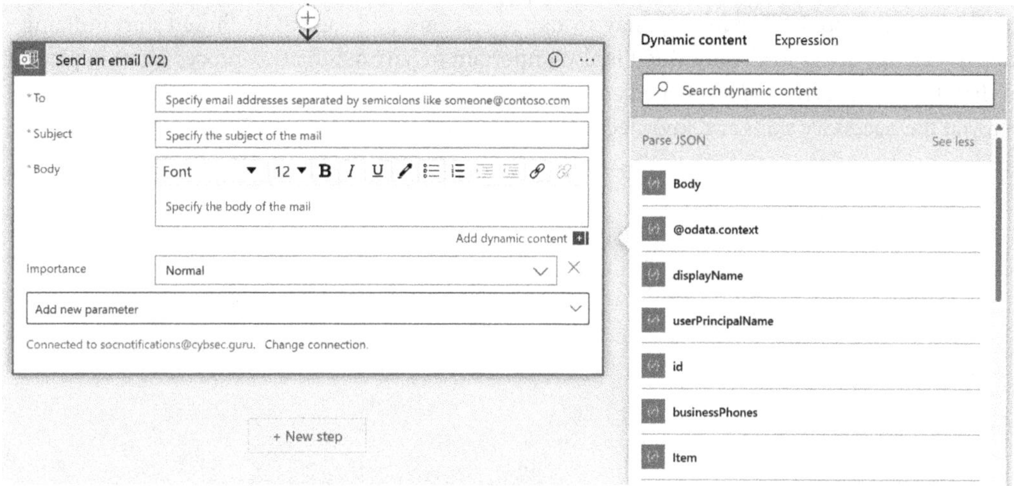

Figure 9.34 – Dynamic content from the Parse JSON action

Mastering the examples demonstrated in this last section will help you to work with Microsoft Sentinel playbooks on a daily basis and create more effective automation. With this, we conclude this chapter on mastering Microsoft Sentinel automation.

Summary

This chapter introduced some common tips and tricks you can utilize when working with Microsoft Sentinel playbooks.

First, we introduced how to work with dynamic content and expressions and how to utilize the power of expressions to get data that is not exposed initially with dynamic content. This provides an easy and fast way to work with the data we get from any action.

After that, we focused on my favorite action – **HTTP**. The **HTTP** action allows you to connect to any service that runs on an API and functionalities that may not be available with native actions. In most cases, this will be when a native action doesn't support using a service principal or managed identity for authentication, which we can utilize when using the **HTTP** action.

We finished the chapter by introducing a few actions that we didn't use in examples but will make your day-to-day operations easier when working with Microsoft Sentinel playbooks. Some of them are action-setting details, such as a timeout or configuring something to continue running after failure or timeouts, as well as specific actions such as **Switch**, **Create HTML table**, **Parse JSON**, and **Compose**.

On mastering Microsoft Sentinel automation, we are wrapping up this book on SOAR. If there is one thing that I would like everyone to remember, it is that automation is one of the most important features that can help SOC analysts in day-to-day operations, and each SOC should start utilizing automation if not doing so already. But it is also important to have automation processes set in place, and that they are clearly communicated and documented. Hopefully, the hands-on examples have given you the necessary skills and knowledge to start exploring automation further.

Index

A

www.packtpub.com

Subscribe to our online digital library for full access to over 7,000 books and videos, as well as industry leading tools to help you plan your personal development and advance your career. For more information, please visit our website.

Why subscribe?

- Spend less time learning and more time coding with practical eBooks and Videos from over 4,000 industry professionals
- Improve your learning with Skill Plans built especially for you
- Get a free eBook or video every month
- Fully searchable for easy access to vital information
- Copy and paste, print, and bookmark content

Did you know that Packt offers eBook versions of every book published, with PDF and ePub files available? You can upgrade to the eBook version at packtpub.com and as a print book customer, you are entitled to a discount on the eBook copy. Get in touch with us at customercare@packtpub.com for more details.

At www.packtpub.com, you can also read a collection of free technical articles, sign up for a range of free newsletters, and receive exclusive discounts and offers on Packt books and eBooks.

Other Books You May Enjoy

If you enjoyed this book, you may be interested in these other books by Packt:

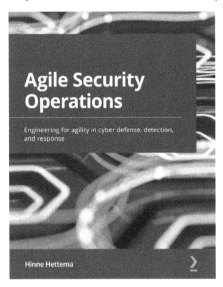

Agile Security Operations

Hinne Hettema

ISBN: 978-1-80181-551-2

- Get acquainted with the changing landscape of security operations
- Understand how to sense an attacker's motives and capabilities
- Grasp key concepts of the kill chain, the ATT framework, and the Cynefin framework
- Get to grips with designing and developing a defensible security architecture
- Explore detection and response engineering
- Overcome challenges in measuring the security posture
- Derive and communicate business values through security operations
- Discover ways to implement security as part of development and business operations

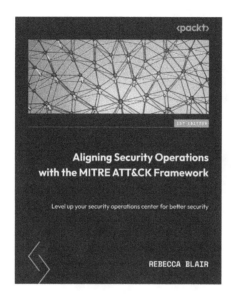

Aligning Security Operations with the MITRE ATT Framework

Rebecca Blair

ISBN: 978-1-80461-426-6

- Get a deeper understanding of the Mitre ATT Framework
- Avoid common implementation mistakes and provide maximum value
- Create efficient detections to align with the framework
- Implement continuous improvements on detections and review ATT mapping
- Discover how to optimize SOC environments with automation
- Review different threat models and their use cases

Packt is searching for authors like you

If you're interested in becoming an author for Packt, please visit `authors.packtpub.com` and apply today. We have worked with thousands of developers and tech professionals, just like you, to help them share their insight with the global tech community. You can make a general application, apply for a specific hot topic that we are recruiting an author for, or submit your own idea.

Share your thoughts

Now you've finished *Security Orchestration, Automation and Response for Security Analysts*, we'd love to hear your thoughts! Scan the QR code below to go straight to the Amazon review page for this book and share your feedback or leave a review on the site that you purchased it from.

`https://packt.link/r/1803242914`

Your review is important to us and the tech community and will help us make sure we're delivering excellent quality content.

Download a free PDF copy of this book

Thanks for purchasing this book!

Do you like to read on the go but are unable to carry your print books everywhere? Is your eBook purchase not compatible with the device of your choice?

Don't worry, now with every Packt book you get a DRM-free PDF version of that book at no cost.

Read anywhere, any place, on any device. Search, copy, and paste code from your favorite technical books directly into your application.

The perks don't stop there, you can get exclusive access to discounts, newsletters, and great free content in your inbox daily

Follow these simple steps to get the benefits:

1. Scan the QR code or visit the link below

https://packt.link/free-ebook/9781803242910

2. Submit your proof of purchase
3. That's it! We'll send your free PDF and other benefits to your email directly

www.ingramcontent.com/pod-product-compliance
Lightning Source LLC
Chambersburg PA
CBHW062059050326
40690CB00016B/3151